Resource Devastation on Native American Lands

Bruce E. Johansen

Resource Devastation on Native American Lands

Toxic Earth, Poisoned People

Bruce E. Johansen
University of Nebraska at Omaha
Omaha, NE, USA

ISBN 978-3-031-21895-8   ISBN 978-3-031-21896-5 (eBook)
https://doi.org/10.1007/978-3-031-21896-5

© The Editor(s) (if applicable) and The Author(s), under exclusive license to Springer Nature Switzerland AG 2023
This work is subject to copyright. All rights are solely and exclusively licensed by the Publisher, whether the whole or part of the material is concerned, specifically the rights of translation, reprinting, reuse of illustrations, recitation, broadcasting, reproduction on microfilms or in any other physical way, and transmission or information storage and retrieval, electronic adaptation, computer software, or by similar or dissimilar methodology now known or hereafter developed.
The use of general descriptive names, registered names, trademarks, service marks, etc. in this publication does not imply, even in the absence of a specific statement, that such names are exempt from the relevant protective laws and regulations and therefore free for general use.
The publisher, the authors, and the editors are safe to assume that the advice and information in this book are believed to be true and accurate at the date of publication. Neither the publisher nor the authors or the editors give a warranty, expressed or implied, with respect to the material contained herein or for any errors or omissions that may have been made. The publisher remains neutral with regard to jurisdictional claims in published maps and institutional affiliations.

This Springer imprint is published by the registered company Springer Nature Switzerland AG
The registered company address is: Gewerbestrasse 11, 6330 Cham, Switzerland

Acknowledgements

Editor Niko Chtouris, of Springer, in Frankfurt, Germany: profound advisor, and a great editor. I wish I had met him a few decades ago; Prof. Barbara Mann, Honors Professor at the University of Toledo, friend and colleague for almost 30 years, always a no-nonsense editor and advisor; as well as Joy Porter, professor of history at Hull University in the United Kingdom; Yvonne Schwark-Reiber, also at Springer, for very good advice on a manuscript turning into a book; Everyone else at Springer, an extraordinary group of editors, writers, proofreaders, members of the production crew, and executives, as well as everyone else who is involved in the vital task of creating, maintaining, and diffusing information in the task of publishing; Pat Keiffer, ("Nana Pat"), my wife, who tends the hearth in our home and provides an eagle's eye for typos and errors of commission and omission on manuscripts; she has also been invaluable to our growing and maturing family in and near Omaha; to hundreds of Native people around the USA, Canada, and Nunavut (Baffin Island), who have taught me, and sometimes fed and housed me with both information and very necessary facts and opinions on the path that provided the factual backbone for this book. Some of the information is more than 50 years old and has been continually updated by travel, surface mail, and Internet. Many of the situations described in the book have improved marginally, but never quickly enough. The struggle continues. Everyone has taught me (and many others) what is important in life.

Contents

1	**Introduction**		1
	References		15
2	**Akwesasne: Land of the Toxic Turtles**		17
	References		34
3	**The Deadly Yellow Powder**		37
	3.1	The Human Price of Uranium among Native American Peoples	37
	3.2	The Scope of Uranium Mining on Native Lands	39
	3.3	Uranium from Russia: An Alarm for U.S. Native Peoples?	40
	3.4	Navajo Uranium Fuels U.S. Arsenal	42
	3.5	Radioactive from the Inside Out	43
	3.6	Opposition to Uranium Mining	45
	3.7	"Original Instructions," Uranium Mining, and Navajo Cultural Values	45
	3.8	The Largest Uranium Spill in the United States	46
	3.9	The Struggle for Compensation	48
	3.10	Delays in Compensation	49
	3.11	Cleanup Planned at a Few Spill Sites after Three Decades	50
	3.12	"Nobody Told Us It Was Unsafe"	53
	3.13	The Laguna Pueblo and Anaconda's Jackpile Uranium Mine	54
	3.14	A Child Nearly Burns to Death	56
	3.15	Decades of Cleanup Plans	57
	3.16	Dene Decimated by Uranium Mining and the "Money Rock"	58
	3.17	"The Incurable Disease"	59
	3.18	The Conflict Continues	60
	3.19	Plans to Blast an Alaskan Harbor with Nuclear Bombs	61
	3.20	The Point Hope Eskimos: An Atomic Harbor and a Nuclear Dump as a Neighbor	62
	3.21	Nuclear Boosterism	62

vii

3.22	An "Overture to the New Era"?	64
3.23	Radiation Tests' Effects on Eskimos	64
3.24	Cancer Rates Rise	65
3.25	The Prairie Island Indian Community and Nuclear Waste	66
3.26	The Western Shoshone: "The Most-Bombed Nation of Earth"	67
3.27	More Tests on Indian Land	68
3.28	Washington State's Yakamas and Hanford's Radioactive Legacy	68
3.29	On the Road to Hell, the Most Toxic Lies Go Undisclosed	70
3.30	The Dene Extend Profound Condolences to the Victims in Hiroshima and Nagasaki	71
References		73

4 Showers of Pig Feces: A Neighborly Stench ... 77

4.1	Money Oinks at the State Capitol	77
4.2	Wearing the Stench	78
4.3	A Stench that Smells like a Month-Old Decomposing Human Body	79
4.4	So Much for Free Speech	80
4.5	More Health Problems Related to Hog Waste	80
4.6	Pig Manure, Race, Class, and Corporate Control	82
4.7	"When Pigs Fly"	83
4.8	Now Comes a Stampede of Chickens and Turkeys	85
4.9	Hogs Now Run Second to Poultry	86
References		87

5 An Ice World Melts ... 89

5.1	Sweating in Iqaluit	89
5.2	Warming in the Arctic and Antarctic: How High Is Spectacular?	90
5.3	When Is it "Weather?" When Is It "Climate?"	91
5.4	Vast New Lakes Created	91
5.5	An Out-of-Season Swarm of Flies	92
5.6	Climate and Cultural Change	93
5.7	Sea Ice Is in Rapid Decline	94
5.8	Food Insecurity and Grocery Stores	95
5.9	"We Have Never Seen Anything Like This. It's Scary, *Very* Scary"	96
5.10	Worst Fears Realized	100
5.11	Land of Melting Ice and Burning Tundra	101
5.12	Warming's Pervasive Effects	102
5.13	A Winter Without Walrus	103
5.14	An Oral History of a Melting World	104
5.15	The Inuit World Turned Upside down	105
5.16	The Arctic Ocean Is Acidifying	106
5.17	Alaskan Villages Fall to Encroaching Seas	106

5.18	Moving Beyond a Point of No Return	108
5.19	Salmon Decline in Warming Waters	109
5.20	Climate Change and Industrial Development	111
References		111

6 The Inuit (and Others): If it Swims, It's Probably Poisonous 115

6.1	"We Feel Like an Endangered Species"	116
6.2	A Poisoned World	117
6.3	How Persistent Organic Pollutants (POPs) Work	117
6.4	Inuit Infants: "A Living Test Tube for Immunologists"	119
6.5	Socioeconomic Changes in Nunavut	121
6.6	Toxic Pollution of Inuit Land by the Military	122
6.7	Toxic Contamination of Traditional Inuit Foods in Alaska and Russia	124
6.8	Seward, Alaska: Don't Eat the Reindeer	125
6.9	Deformed Babies, Mercury, and the Grassy Narrows First Nation in Ontario	126
6.10	Too Few Infant Boys? Blame Estrogen-Blocking Chemicals	126
6.11	The Ouje-Bougoumou Cree Endure Severe Metal Contamination	127
6.12	Colville Tribes Resist Pollution of the Columbia River	129
6.13	The Penobscots Endure Organochlorine Contamination	130
6.14	Oregon Tribes Fight Pollution in and near Portland	131
6.15	Coeur d'Alene Demand Cleanup of Mining Waste in Idaho	131
6.16	The Ramapough: Suffering Ford Motor's PCBs, Heavy Metals, Freon, Arsenic, and Lead	132
6.17	The Yaquis' Borders Don't Stop Pesticide Contamination	133
6.18	The Huicholes Live with Pesticides Around the Clock	136
6.19	Pregnant Inuit Women more Exposed to "the New PCBs" than Other Canadians	139
6.20	How PCBs Move	141
6.21	The Scope of Knowledge About Effects of Contamination Grows	144
6.22	"Do we Become Farmers?"	144
6.23	Bad News on Whales near the Faroe Islands	146
6.24	Mercury Also Damages Childrens' IQ	147
References		147

7 Alberta's Moonscape: If This Sounds Apocalyptic, It Is 151

7.1	Many Pipelines, Not Much Publicity	151
7.2	The Land Is Sacred, but Also a Battlefield	152
7.3	A Perpetual Battle	153
7.4	Tar Sands' Devastation in Northern Alberta	154
7.5	Promotion and Erosion of Support for Tar Sands Oil	154
7.6	Tar Sands and Strip Mining	155

7.7	Tar Sands Mining as the "Fuse to the Biggest Carbon Bomb on the Planet"..	155
7.8	A Signature Environmental Issue............................	156
7.9	Tar Sands as Junk Energy................................	157
7.10	Strip Mining a Moonscape................................	158
7.11	Tar Sands and Climate Change: "If This Sounds Apocalyptic, It Is"..	158
7.12	Tar Sands and Oil Spills.................................	159
7.13	On the Ground in Tar Sands Country........................	160
7.14	Oil on the Water Table..................................	161
7.15	Tribes and Nations Forge Alliances.........................	162
7.16	Native Peoples Unite against the Keystone XL.................	165
7.17	A Blockade at Pine Ridge................................	166
7.18	Another Blockade by the Nez Perce.........................	167
7.19	Protests Spread.......................................	168
7.20	More Protests, More Arrests..............................	170
7.21	Invoking the Fort Laramie Treaties.........................	172
7.22	Shipping Oil: And Spilling It—By Rail......................	172
7.23	The Quinault Resist Oil Transport..........................	173
7.24	More Resistance to Fracking..............................	174
7.25	Fracking and Earthquakes................................	177
7.26	Fracking Linked to Water Pollution.........................	178
7.27	Oil at Fort Berthold: "The Water is Dead and It Is Lethal".......	179
7.28	Protests of Pipelines across Canada.........................	181
References...		185

8 Mining: Tearing at Mother's Breast............................. 191
8.1	The World's Largest Salmon Run Vs. the World's Biggest Gold and Copper Mine.................................	191
8.2	Common Cause at Standing Rock..........................	197
8.3	Oil Ports, Pipelines, Salmon, Dams, Native Peoples: And Greenhouse Gases.....................................	200
8.4	Reorientation Respecting Humans' Relationship Toward the Natural World..	202
8.5	Quapaw: Too Toxic to Clean up...........................	204
8.6	Coal: Pacific Northwest Tribes Protest Transport...............	206
8.7	Coal Transport and the Global Greenhouse Gas Load...........	211
8.8	The Lubicon Cree: Land Rights and Resource Exploitation......	213
8.9	Logging on Lubicon Land...............................	215
8.10	Conflicts over Resources Continue.........................	215
8.11	The Moapa Paiute: Goodbye Toxic Ash: Solar in, Coal Power out..	216
8.12	Montana's Gros Ventre and Assiniboine: Gold Mining and Cyanide Poisoning....................................	219
8.13	"Like Watching Our Ancestors Die"........................	220

8.14	The Mine Leaks, and Expands	221
8.15	Streams Smell of Rotten Eggs	222
8.16	The Huichol (Wixáritari), a Sacred Site, and Silver Mining	223
References		224

Chapter 1
Introduction

By the beginning of the third millennium on the Christian calendar, observers of humanity's relationship to the Earth began to realize that a very important breach of prior conditions had been reached. Humankind had taken commanding control over the Earth and its future. What had been described as natural disasters became human-aided, whether climate-related, the spread of pollution, and many other manifestations of environmental dysfunction. People who observe Earth's condition now came to speak of the Anthropocene, when human change upon the Earth dominates nature. The Earth, for example, has always experienced climate change, and disasters, but never before with the speed and endurance of human-aided cycles.

At the same time, more people began to understand that the planet's epical problems cannot be solved by almost 200 countries on Earth in isolation, and most assuredly not in bloody war with each other. To solve global warming, for example, the peoples of the Earth also must learn how to end nationalism as a motor of war, as well as we move away from combustion of fossil fuels. This will not be at all easy, since it goes against nearly every inclination of human beings' behavior with each other since we discovered the use of weapons and language. Ever since the conflict was waged with sticks and stones, the destructive power of nuclear weapons has grown to the point where no one dares use them except a deranged, unhinged leader. At the same time, in addition to the existential threat of nuclear apocalypse, combustion of fossil fuels steadily warms the atmosphere, presenting another prospective apocalypse that demands solution very quickly on nature's clock. Who is ready to face the future framed in these terms? How do we learn to satisfy our material needs without ruining our home? Perhaps we begin this journey of reformation by studying the histories of the world's indigenous peoples instead of forcing them to minimal survival at the dangerous edges of an economic system that irradiates their bodies and crushes their souls. We must learn rather than exploit, a massive challenge to a human race with a history of forcing its indigenous wisdom bearers into poverty since *human* expansion around the world has been characterized by *inhumanity*.

We present here a sordid, wrenching description of indigenous peoples and their homelands driven close to extinction by industrial humankind's drive to exploit the Earth's natural resources for use by an industrial state that grows pell-mell, driven by the human will to profit from others' lives and labor.

"With our fertilizer plants, we fix more nitrogen than all [other] terrestrial eco-systems combined [as well as growing dead zones in lakes and oceans], with our plows and bulldozers, we move around more earth than all the world's rivers and streams. In terms of biomass, the numbers are staggering. People now outweigh wild mammals by ratio of more 8 to 1. Add in our domesticated animals (mostly cows and pigs, and the ratio is more than 23 to 1....All sorts of catastrophes straddle the line between man and nature" (Kolbert 2021, 16). In search of oil by "fracking," earthquakes shake areas where the natural observations of geology do not explain them. Intense pandemics such as COVID-19 spread and mutate around the world with a speed heretofore unknown with the crowding of human populations and the speed and ubiquity of jet aircraft. People have always endured plagues, and have raced to stay ahead of their ability to mutate even as we chase them with medicine, even as they sweep the world with unprecedented speed. Within a month of its first detection in Wuhan, China, COVID-19 had swept into at least 26 countries (Kolbert 2021, 16). In another 10 months, a million people had died, as variants (mutations) were competing with vaccines for dominance. Within another few. Months, the death toll from COVID had exceeded one million in the United States *alone*. These are large numbers, but not so many in proportion to the 8 billion or more people who populate the Earth. No doubt a worldwide nuclear war sparked by runaway nationalism would kill more than that, while blighting humanity's future with deadly radiation.

The unique aspect of such a war would be its human creation. Nuclear power begins with uranium most often mined on the lands allotted to Native American peoples who have been suffering its effects for almost a century by now. This book visits only some of uranium's killing fields.

Another human-made catastrophe that often victimizes human beings who are least able to cope with it has been the growing, world-girding effects of climate change. These changes often victimize indigenous peoples more often and with greater intensity than others.

This work examines environmental problems stemming from resource exploitation on Native lands from Nunavut and the Canadian Arctic southward through the United States. It is mainly contemporary, but all of these problems have historical roots. The work focuses on Native resistance to resource colonization that is addressing the many environmental crises across Native North America that have become a plague upon many indigenous peoples.

Pollution nearly inevitably follows resource exploitation, Native peoples in the United States today often live on ruined, exhausted land, suffering toxic consequences. *As of June 5, 2014, 532 (about 40%) of 1322 Superfund sites in the United States were in Indian country* (Hansen 2014). *This book examines a wide variety of serious environmental issues affecting scores of Native peoples in Alaska, Canada, and in the United States of America.*

1 Introduction

Many reservation residents are dealing with illnesses due to the use of their lands for several decades as industrial dumps and mine sites. This work examines the acute effects of exposure to dioxins, PCBs, and other persistent organic pollutants in several Native communities, most acutely in the Arctic, where Native consumption of sea life (their traditional diet) has been curtailed and Inuit mothers are sometimes warned not to breastfeed their infants. Mothers' milk may be toxic. This book also examines the present and prospective effects of global warming (again, most acute in Arctic areas of Canada, and Alaska) on Native peoples. Salmon, for example, are imperiled by warming water. The Arctic, which looks so pristine to the untutored eye, is also experiencing the world's most rapid rate of climate change. An Inuit culture based on ice is melting. and many Inuit hunters have been injured or killed by falling through thin ice.

I have been writing about indigenous peoples' resistance to changes in their homelands for 50 years, starting with a critique of Earth Day *vis a vis* tribal people bombed and poisoned by chemicals in the Vietnam war as editor of my college newspaper at the University of Washington in 1970. I also began by covering Puget Sound Indian fishing rights as an intern at the *Seattle Times*. Since then, I have produced copious books, magazine and academic-journal articles, as well as newspaper articles describing the struggle to change how to change energy production, popular perceptions, political practices, technology, et al. to accommodate sustainable ways of life before it is too late.

This pursuit has taken me into several academic fields, including environmental studies, the histories and legacies of indigenous peoples with the sciences and politics that comprise today's debates about climate change. I am one of very few academic researchers and writers who can summarize scientific and technological knowledge in a context of indigenous peoples' histories and cultures into a narrative that can be understood and appreciated by people in many fields of study and levels of endeavor. All of this has been undertaken within a context of indigenous environmental knowledge.

I was working on my first book, *Wasi'chu: The Continuing Indian Wars*, in 1976 when a Navajo elder, Emma Yazzie, of the Coalition for Navajo Liberation introduced me to uranium and coal mining in Navajo Country. Women, especially elders, have been prominent in modern-day Native social and political movements. Yazzie, who was more than 70 years of age when I first met her (I was 26 years old at that time, in 1976), Yazzie became an important Navajo leader against coal strip mining as her herds of sheep were devastated by mining pollution (Johansen and Maestas 1979, 143–144).

Yazzie led Roberto Maestas and me to the floor of a strip mine near her hogan, disregarding the orders of mine security, to watch draglines at work, saying: "Don't go alone. They'll turn you away. They're afraid of me!" (Johansen and Maestas 1979, 145). Yazzie showed her disgust at the mining by removing surveyors' sticks and dumping them on supervisors' desks.

Over the years, Yazzie had found herself living between coal strip mines and the Four Corners coal-fired electrical plant, which (while it was operating, until November 18, 2019) put out a polluting plume so large that NASA astronauts observed it from Earth orbit. Yazzie's sheep had, by 1976, become small, skinny, and sickly, with gray wool. The Four Corners plant exported power to Los Angeles, Phoenix, Las Vegas, Nevada, and other cities in the region. Yazzie's Hogan had no electricity. The transmission lines buzzed above Emma Yazzie's hogan, which contained not so much as a single light bulb.

When she gave me this tour, Yazzie was standing under the largest methane "hot spot" in North America. This area on and near the Navajo Nation, which stands out on satellite photographs taken between 2003 and 2009, had been created by coal mining and coal-fired power generation from the Four Corners Power Plant on the reservation and the San Juan Power Plant nearby. Some of this power generation later was shut down to avoid expensive pollution controls. The methane "hot spot," which covers about 2500 square miles, roughly half the area of Connecticut not far from the intersecting borders of Colorado, Arizona, New Mexico, and Utah, was described in the October 16, 2014 issue of *Geophysical Research Letters* (Kort et al. 2014). Before it closed, the Four Corners methane plume by itself comprised 10% of U.S. methane emissions, according U.S. Environmental Protection Agency estimates (Minard 2014).

Yazzie did not know the academic language of resource colonization and exploitation, but she *did* know that the plume of smoke and ash that blew upslope near her hogan stunted the growth of her lambs and stained their coats dirty brown. She called the mine and the power plant a monster and a harbinger of death.

During the 1940s, representatives of the companies that mined some of the uranium which powered the United States' growing nuclear arsenal (and, later, early nuclear power plants) had arrived on the Navajo Nation offering prosperity through mining jobs. Left unstated was the eventual cost of this uranium boom: an epidemic of cancer (especially lung cancer) in a place that had never known such a thing. Until recently, cancer had no name in the Navajo language. Today it rides the winds across the reservation as yellow dust from huge piles of tailings (mining waste). The mining and milling of uranium is now illegal on Navajo land, but a deadly legacy remains.

Uranium mining has been only one part of energy colonization on the Navajo Nation that led the U.S. Energy Department during the 1970s to call it a "national sacrifice area" (Johansen and Maestas 1979, 143, 170). A large part of this sacrifice involved the use of plentiful coal reserves under the Navajo Nation to generate electricity, thrumming along transmission lines to Los Angeles and other cities, over hogans that had no power at all, meanwhile polluting the air that Emma Yazzie and many other Navajos breathed. Navajos continue to battle for a cleaner environment; in 2015, the Environmental Protection Agency announced a $168 million settlement that may remedy part of the pollution over several years around the Four Corners plant that heretofore had ruined Yazzie's flock 40 years before. Neighbors of the plant are still suffering from asthma and other health maladies with inadequate access to health care.

Near the town of Page, the Navajo Generating Station, then the largest coal-fired plant in the U.S. West (It was demolished during December 2020) poured plumes of carbon dioxide, lead, mercury, nitrogen oxide, as well as other metals from three

775-foot stacks. Abraham Lustgarten of ProPublica reported, in 2015, that "Every hour the Navajo's generators spin, the plant spews more climate-warming gases into the atmosphere than almost any other facility in the United States. Alone, it accounts for 29% of Arizona's greenhouse-gas emissions from energy generation. The Navajo station's infernos gobble 15 tons of coal each minute, 24 hours each day, every day." According to the Navajos' government, cancer rates in the area have doubled since the plant began operations four decades ago (Lustgarten 2015). "You are trying to raise your family in this environment, and you realize this is one of the top 10 dirtiest plants in the nation and it's been spewing all this stuff for 40 years," said Nicole Horseherder, a Navajo environmental activist. "Who is going to speak up and say, 'Look, we are paying a huge cost so that the state of Arizona can have its profits, have its taxes, have its electricity, have its water?'" (Lustgarten 2015).

When reservations (or, in Canada, reserves), were assigned to American Indians in the mid-to-late nineteenth century the fossil-fuel age was just beginning. The uses of uranium were largely unknown, and the federal governments of the United States and Canada, as well as corporations, had only the slightest inkling of what mineral and fuel riches lay under the reservations' often-barren soil. By the late twentieth century, however, these riches were being exploited, and many Native nations became virtual resource colonies.

A report from the Federal Trade Commission distributed during October 1975 said that an estimated 16% of the United States' uranium reserves that were recoverable at market prices were on reservation lands; this was about two-thirds of the uranium on land under the legal jurisdiction of the United States federal government. There were, at that time, almost 400 uranium leases on these lands, according to the F.T.C., and between 1 million and 2 million tons of uranium ore a year, about 20% of the national total, was being mined from reservation land.

The toxic legacy of Native North America is pervasive but largely invisible to most of us. Many toxic sites are located in out-of-the-way rural areas largely forgotten by the majority of America that nonetheless supplied its industries with the rudiments of manufacturing for the better part of a century before being closed and cast aside.

Examples abound from sea to shining sea. To cite only one of many, zinc and lead were mined within the jurisdiction of the Quapaw in Picher, Oklahoma, until 1967, when, according to a retrospective by Terri Hansen on the Indian Country Today Media Network, "mining companies abandoned 14,000 mine shafts, 70 million tons of lead-laced tailings, 36 million tons of mill sand and sludge, as well as contaminated water, leaving residents with high lead levels in blood and tissues. Cancers skyrocketed, and 34% of elementary-school students suffered learning disabilities" (Hansen 2013). The area was designated as the Tar Creek Superfund site in 1983, but Picher initially was skipped for remedy and assigned to a special environmental hell: too toxic to clean up. Instead, the federal government offered a rock-bottom buyout that paid people to leave town.

This is not an exceptional case. A detailed catalog of such environmental atrocities would fill a very thick volume. What follows in this book is merely a sampler from across North America. I have added a few that are not on Indian reservations (such as dioxin in Vietnam and pig waste in North Carolina) to illustrate the pervasive nature of these types of pollution.

Travel nearly anywhere in North America, and toxicity has become heritage. In Alaska, the Salt Chuck Mine, a source of copper, gold, palladium, and silver between 1916 and 1941, contaminated the Kasaan (Alaska) harvesting grounds for fish, clams, cockles, crab, and shrimp. For decades, the Native people were unaware that their harvests were saturated with effluvia from mine tailings. Even after the area was declared a Superfund site, Pure Nickel, Inc. sought to reactivate mining in 2012.

To many Native Americans—those with a naturalistic philosophy—mining is the ultimate insult, a rape of Mother Earth transformed into mother lode, thus the number and intensity of protests by indigenous peoples against mining across Turtle Island, the Iroquois name for North America. The mining of uranium is the most notably odious; it is such a natural breach that Navajo cosmology warns against it. Mercantile capitalism has no such qualms about mining. It is, in fact, the very basis of a system that survives and thrives by making and selling things, thus the germ of conflict: Native American homelands contain vast stores of exploitable natural resources that corporations find useful and profitable, often available from the federal government at prices far below market value—or so they were until the "other" started talking back.

The catalog of mining's toxic legacy on Native American lands is vast. Many Native peoples across North America have organized against mining on their homelands, notably in Alaska, where the world's largest sockeye salmon run could be imperiled by the proposed Pebble Mine, potentially the largest gold and copper strip mine on Earth, which could produce as much as 80 billion pounds of copper. A permit for that mine was denied in 2014 by the U.S. Environmental Protection Agency (EPA), but mining companies are challenging the ruling. Which could be reversed following the election of a Republican president in 2024. This might be a re-run of the 2016 election of Donald Trump and his appointment of Scott Pruitt to head the Environmental Protection Agency. Given such a prospect, the Pebble mine (and others) will be back in play.

In Wisconsin, copper, zinc, and sulfide mining are being proposed, and they are being resisted by Native peoples. Silver mining in Mexico is poisoning indigenous children. Coal strip mining is an important issue for Hopis and Navajos, although they have been phasing out dependance on coal, and have begun closing coal-fired electric-generating plants, the best-known of which the Four Corners behemoth.

In California, the Elem Band of Pomo Indians now are suffering elevated levels of mercury in their bodies due to the Sulfur Bank Mine, a superfund site that borders their colony. Hansen (2014) wrote that "nearby Clear Lake is the most mercury-polluted lake in the world, despite the E.P.A.'s spending about $40 million over two decades trying to keep mercury contamination out of the water. Although the E.P.A. has cleaned some of the soil under Pomo roads and houses, pollution still seeps beneath the earthen dam built by the former mine operator, Bradley Mining Co. For years, Bradley Mining has fought the government's efforts to recoup cleanup costs."

Resource colonization has continued for the entirety of United States history. Coal, oil, coal, gas, and other minerals worth $3.5 billion were extracted from Native American lands in the United States during the fiscal year 2011, compared to $2.8 billion worth in the fiscal year 2010, according to a report by the Government Accountability Office (GAO).

In the United States, Native American reservations represent only 2% of the land but hold approximately 20% of the country's fossil fuel reserves, including coal, oil, and natural gas. Together these fuels are worth about $1.5 trillion, according to the Council of Energy Resource Tribes (Osborne 2018). Indian reservations contain almost 30% of the United States' coal reserves west of the Mississippi, 50% of potential uranium reserves, and 20% of known oil and gas reserves—resources worth nearly $1.5 trillion, or $1.5 million per member of Native Nations. Yet 86% of Indian lands with energy or mineral potential remain undeveloped because of Federal control of reservations that keeps Indians from fully capitalizing on their natural resources if they desire.

Meanwhile, most American Indians live in poverty, with an average per capita income of $16,645 (compared to $27,334 for the U.S. population as a whole) and unemployment rates as high as 78% on some reservations (Regan and Anderson 2013, p. 2. This report contains no dates, limiting the usefulness of its statistics. Maps and statistics indicate, without sayings much, that Alaskan native corporations also may be excluded.) Although in the US, Native American reservations represent only 2% of the land, they hold about 20% of coal, oil, and gas reserves of the country. An audit by the Energy Resource Tribes puts the total value of all these fuels together at somewhere around $1.5 trillion.

Reservation governments received some monetary benefit from this activity. Income to Native American reservations for extraction of fossil fuels was about $540 million in the fiscal year 2011, up slightly from about $400 million in 2010. By 2016, the total Native American income from fossil fuel extraction topped $1 trillion, and by 2020 it was estimated at $1.5 trillion. These rough estimates did not include the value of copper, gold, and other hard-rock resources from federal or Indian lands; mining companies are not legally obligated to pay royalties. They pay fees for leases. Resource exploitation of first nations in Canada also has developed along similar lines. In 2014, American Indian revenues from mineral leases in the United States (34,000 owners in 34 tribes and nations) passed $1 billion for the first time, up about $200 million from 2013

Environmental provocations afflicting Native American peoples in the United States (a range of problems equal to those of any Third World nation) range from uranium mining to kitty litter for the Navajos to the devastation wrought by dioxin, PCBs, and other pollutants on the agricultural economy of the Akwesasne Mohawk reservation in northernmost New York State. As with the Akwesasne Mohawks, some of the most serious problems span international borders. The Yaquis, whose homelands span the U.S.–Mexican border, have been afflicted with some of the same pesticides as the Mohawks on the U.S.—Canadian border.

Some of the environmental problems faced by indigenous peoples in the United States strain one's sense of credulity. Witness the Eskimos of Point Hope, Alaska, who have learned that parts of their land once had been once been eyed as a harbor for a harbor to be created by exploding nuclear weapons. Although the harbor was not exploded open after protests by Point Hope Eskimos and white environmental activists, the Point Hope Eskimos still found themselves dealing with nuclear waste that they had not requested. Other Alaskan Eskimos found reindeer inedible, their bodies polluted by several heavy metals. The Western Shoshone of Nevada called themselves "the most bombed nation on Earth," after a nearby nuclear test range and a since-canceled proposal at Yucca Mountain to open a national waste-uranium repository. The U.S. government seemed to be acting as if no one lived in the area.

Canada, which often tells outsiders how humane their civil rights record has been regarding indigenous peoples, has nevertheless become a major source of conflict and contamination. The Innu of Labrador have found themselves afflicted by aluminum smelting and sulfide mining, Overhead, the previously quiet skies at times have been filled with noise from military aircraft.

A remote location in Canada is no guarantee against intense resource exploitation. In northern Alberta, among the Lubicon Cree, lands were so inaccessible that treaty surveyors completely missed them. By 2000, however, roads had opened their lands to logging and massive test oil drilling. The Cree of Quebec found their homelands a target of dam building near James Bay that contaminated large areas with toxic methylmercury as well as other pollutants. Mining has decimated the Dene who live in Canada's Northwest Territories much as it has spread cancerous illnesses among the Navajo in New Mexico and Arizona.

The pesticides can be found in many Native areas within the United States, and their use spread even to Vietnam during the war there. Less than a decade of intensive use was enough to produce birth defects and many other problems for generations. Now, more than 50 years later, thousands of Vietnamese face lives crippled by deformations and many other maladies from dioxin (Agent Orange). The same medical conditions also have been diagnosed in U.S. veterans (as well as *their* children) and generations after that. The effects of severe problems traced to Agent Orange will probably be detectable. For five to ten more generations after the present. As one of many "persistent organic pollutants" (POPs), dioxin is nearly impossible to cleanse from the bodies of human beings and other animals. Some of the environmental problems faced by indigenous peoples in the United States strain one's sense of credulity. Witness the Eskimos of Point Hope, Alaska, who have learned that their land had once been proposed as the site of a new harbor to be created with nuclear weapons. The harbor was never created, but the Point Hope Eskimos still found themselves hosting uninvited nuclear waste. Other Alaskan Eskimos have found their reindeer rendered inedible, polluted with a number of heavy metals. The Western Shoshone of Nevada have come to call themselves "The most bombed nation on Earth," a reference to a neighboring test range for nuclear weapons and (more recently) a now-stalled proposal to open a national waste-uranium repository at Yucca Mountain, which was shut down after several years of protests.

The Innu of Labrador have been afflicted with sulfide mining, aluminum smelting, and noise pollution from squadrons of military aircraft. Some of the most intense resource exploitation in Canada takes place in remote locations, such as among the Lubicon Cree of northern Alberta, whose lands were so inaccessible that in 1900 that treaty map makers completely missed them. Today, roads have opened their lands to massive oil drilling and logging. The lands of the Cree in Quebec have been scarred by widespread dam building near James Bay that has contaminated large areas with toxic methyl mercury, as well as other pollutants. Uranium mining has devastated the Dene in Canada's Northwest Territories much as it has ravaged the Navajo in the Southwestern United States. Indigenous environmental issues in Canada span the range of resources—from hydropower, to diamonds, uranium, gold, silver, and sulfide, aluminum, oil, and natural gas.

Several first nations in Alberta been scarred by the development of oil from tar sands.

Mexico's indigenous peoples have been dealing with a wide variety of environmental issues, from the Mayas of Chiapas in the South, (who have been protesting the intrusion of roads, dams, logging, and hydropower dams under the Zapitista banner), to the Huicholes Indians of Sinaloa, Baja California, who harvest herbicide-laced tobacco. The workers and their families not only experience toxicity on the job; they also live in the fields. Elsewhere in Mexico, The world's largest producer of silver has been ordered to cut production because children near its facilities were found to have unhealthy levels of lead in their bodies. Near Mexico City, in the State of Morelos, at the birthplaces of the Mexica (Aztec) god Quetzalcoatl as well as Emilio Zapata, local indigenous peoples took to the streets to resist the construction of a golf course that would have taken much of their water in a historically dry region. The image of Zapata was invoked by the protesters, who eventually defeated the proposed golf course.

A count of all such environmental atrocities could fill might fill a small encyclopedia; this chapter describes just a few. It follows environmental atrocities in the south of Baffin Island, in the south of Baffin Island and the Mohawk reservation of the very Northmost New York States, with brief descriptions of Mexico's Huicholes. The Arctic is hardly as pristine as popular imagination has it, as polychlorinated biphenyls (PCBs) and dioxins reach its upper atmosphere and reach the ground as poisoned snows. A student of history might recall dioxins as a major component of a powerful herbicide used in Vietnam used to kill forests that also produced major deformations of young Vietnamese, as it spread through generations. Anyone who has used RoundUp to kill weeds and produce a weed-free "perfect" lawn has been using a milder form of dioxins.

Nunavut, part of the Inuit Nation, a semi-sovereign territory, receives its chemicals, swept northward by prevailing winds from industries in the United States and Canada, until they were outlawed by the Stockholm Convention. Unfortunately, the chemicals are very difficult to remove from the food chains that affect marine mammals such as whales and seals. The sea mammals are a favored food of the Inuit, who receive an even higher dose when they are eaten by humans by a process known as biomagnification, which increases their potency at each step up the food chain. Small fish eat even smaller sea creatures, which have been toxified at a relatively low dose of these chemicals. The fish are eaten by a whale or a seal at a higher level, and a human being at an even higher level. The rise in potency occurs at roughly exponential levels; that is: Very small fish = 1; small fish 2 (2 × 1); whale = 4 (2 × 2); human being = 16 (4 × 4).

Inuit mothers have been advised not to breastfeed their infants because they become the next step up the food chain – 16 × 16 = (256). At Akwesasne, as well as among the Inuit, the larger sea animals have been laced with seriously high levels of PCBs and dioxins. The fish are inedible, and they have been told to avoid eating anything from the St. Lawrence River), long their main source of protein.

Both the Inuit and the Mohawks have organized politically to restore an Aquatic environment that will not poison them. The Inuit, in particular, have had an active role to outlaw the use of these toxins with the Stockholm Convention.

Following negotiation of the Stockholm Convention that outlaws most of the "Dirty Dozen" persistent organic pollutants (POPs), Inuit activist Sheila Watt-Cloutier brought tears to the eyes of some delegates when she expressed gratitude on behalf of the Inuit. The treaty, she said, had "brought us an important step closer to fulfilling the basic human right of every person to live in a world free of toxic contamination. For Inuit and indigenous peoples, this means not only a healthy and secure environment, but also the survival of a people. For that I am grateful. *Nakurmiik.* Thank you".

It is a long road back, however. As pointed out above, these chemicals are often very difficult to expel from their bodies. The battle to rid their food chains of them may require decades.

In addition to being rooted in their homelands, Native Americans maintain historical, spiritual bonds with the land that fosters attention to environmental threats. Longtime fishing rights activist Billy Frank Jr. said that this connection places protection of environment and its role in sustaining human and all other life "at the top of our priority list" (Russo 2012, 235). "When we say the Okanagan [native] word for 'ourselves,'" said Jeanette Armstrong, "we are actually saying the ones [that] are 'dream' and 'land' together" (Grossman 2012, 38). Chief Willie Charlie has said, "Mother Earth is crying, and we need to pay attention to what she is saying" (Grossman 2012, 45–46).

Activist and author Kurt Russo, who works with the Native American Land Conservancy, commented that "our courage to acknowledge this crisis, our conviction to stand up for unborn generations, our connection to nature and, through nature, to each other, and our resilience—as a family, as a species, as peoples—will determine whether we hear her cry for the agony of extinction or stand idly by and bear witness to a great dying" (Russo 2012, 236).

"Indigenous communities are, in general, in a unique position given their history and knowledge to understand and respond to the crisis," commented Russo. "[They] are, in general, more informed and engaged than the majority of Americans or their political and corporate leaders.... As place-based communities of inter-related families with historical consciousness, indigenous peoples are also more resilient, and thus able to face, rather than deflect or deny, the true magnitude of the crisis [that] will have long-lasting and potentially catastrophic consequences for every life-form and human community" (Russo 2012, 234).

On September 29, 2011, Daniel Yuchi member of the Muscogee Nation of Oklahoma and a professor at Haskell Indian Nations University in Lawrence, Kansas, speaking at the Center of the American West of the University of Colorado, Boulder, said that, until recently, Native people were members of tribes, not nation-states, with a relationship to nature that defined species on which people depended for survival in familial terms as relatives, not as exploitable resources. The bison on the Great Plains, salmon among the Coast Salish, and corn across much of today's North America (Turtle Island) were "the central relatives we acknowledged" (Berry 2011).

Indigenous activism against the development of extractive industry that destroys environmental integrity is not new in this area. Indigenous people have long historical experience with resource exploitation that works against a spiritual ethos that invests spiritual integrity in everything that is natural. European religions usually restrict their effects to humanity, many Native Americans interpret their worldview to apply to "all my relations" to mean *all* of nature—animate and not—even the rocks on which we walk. This respect for nature is fundamental and enduring and at the root of traditional Native American responses to economic development. Definitions of "balance" are couched in this context, counterpoising protection of "mother earth" with European immigrants' ethos that seeks a "mother lode," from *Genesis* ("Go forward, multiply, and subdue the Earth").

Naomi Klein, in *This Changes Everything: Capitalism and the Climate* (2014, 183) writes of industrialists who "view … nature as a bottomless vending machine."

The non-Native world is slowly recognizing the toll that its appetites for natural resources wreak on indigenous peoples. In February 2013, the United Nations UN Special Rapporteur on the Rights of Indigenous Peoples James Anaya issued a request "for information on extractive and energy industries in or near indigenous territories … and the human rights issues that these industries generate … such as mining, petroleum or gas projects and energy production, including hydroelectric projects." A report to be compiled by Anaya's office "will identify good practices for the industries to use in order to avoid or overcome … human rights violations." This report was compiled because, according to the Indian Country Today Media Network, "sovereign Indian nations, particularly in the American West, continue to be significantly impacted by radioactive and other hazardous wastes because of the proximity of nuclear test sites, uranium mines, power plants, toxic waste dumps and extractive industries" (Toensing 2013).

In addition to corporations, the U.S. military has been a source of pollution in Indian Country, according to Gregory Hooks and Chad L. Smith, who coauthored "The Treadmill of Destruction: National Sacrifice Areas and Native Americans" in the *American Sociological Review*. The study found "that much of the disproportional exposure of Native Americans to environmental dangers throughout the 20th century was the result of militarism, rather than economic competition. And it shows that historically coercive governmental policies in locating Indian reservations are a major factor in determining their exposure" (Hooks and Smith 2004, 558).

In Chap. 2, "Akwesasne: Land of the Toxic Turtles," our account takes us to a Native American community where, for several centuries of human occupancy, the site the Mohawks call Akwesasne was a natural wonderland: well-watered, thickly forested with white pine, oak, hickory and ash; home to deer, elk, and other game animals. The rich soil in the bottomlands of a valley into which several rivers flowed allowed farming to flourish. The very name that the Akwesasne Mohawks gave their territory about 1755 testifies to the bounty of the land. "Akwesasne" in the Mohawk language means "Land Where the Partridge Drums," after the distinct sound that a male ruffled grouse (partridge) makes during its courtship rituals. Lying at the confluence of the Saint Lawrence, Saint Regis, Racquette, Grass and Salmon rivers, Akwesasne, until recent times, also provided its human occupants with large runs of sturgeon, bass, and walleye pike.

Within roughly less than a century, this land of natural wonders has become a place where one cannot eat local fish and game, because their flesh now is laced at toxic levels with PCBs and other carcinogens. In some places, one cannot drink the water, for the same reason. In parts of Akwesasne, residents have been told to plow under their gardens, and to have mothers' breast milk tested for contamination. In place of sustaining rivers and a land to which the Mohawks still offer thanksgiving prayers, capitalism has offered incinerators and dumps for medical and industrial waste. Akwesasne, which straddles New York State's border with Quebec and Ontario, has become the most polluted Native reserve in Canada, and a number-one toxic site in the U.S. Environmental Protection Agency's "Superfund" list of sites badly needing cleanup.

Within the living memory of many people at Akwesasne, "The Land Where the Partridge Drums," has inherited the toxicological consequences of General Motors' waste lagoons in which animals have been found with levels of PCBs in their fat that qualifies them as toxic waste under U.S. Environmental Protection Agency guidelines. Akwesasne has become riskier to human health than most urban areas, a place where any partridge still living may be more concerned about its heartbeat rather than its drumbeat.

Chapter 3, "The Deadly Yellow Powder," notes that uranium mined from Native American lands supplied a substantial proportion of the fuel for early nuclear power plants as well as the U.S. nuclear arsenal. By the 1970s, many of the early miners were dying of lung cancer, most notably among the Navajo and the Dene of the Canadian Northwest Territories. In Washington State, nuclear waste from the Hanford plants afflicted the Yakamas. Uranium mining also has caused a plague of cancer among the Laguna Pueblo. The Navajos organized to stop uranium mining and milling, and today both are illegal there—but only after several hundred people had died.

Chapter 4, "Showers of Pig Feces" describes the environmental price paid where pork is mass produced. Massive pig farms on an industrial scale produce lakes of swine waste and imperil the air and water of disproportionally black and poor people who endure frequent dousings of pig manure sprayed around their homes to fertilize nearby fields in, as Lily Kuo wrote in *Quartz* (2015), "a fine mist of liquefied feces collects on their houses and cars, attracting swarms of flies." "The poor people, they literally get shit on," said Kemp Burdette, who advocates better water quality in North Carolina's Cape Fear River Watch.

In the language of agricultural bureaucracy, these gigantic pig farms are known as Swine CAFOs (Confined Animal Feeding Operations). The School of Public Health at the University of North Carolina--Chapel Hill reported that "these pig farms are responsible for both air and water pollution, mostly due to the vast manure lagoons they create to hold the enormous amount of waste from the thousands of pigs being raised for food". North Carolina is one of several states in which the number of pigs exceeds the human population, and they produce several times more waste per pig than human beings. No one seems to know exactly how much more (estimates range from 2 to 14 times).

Global warming affects everyone, some more rapidly and intensely than others. *Chapter 5, "An Ice World Melts,"* surveys the Canadian and Alaskan Arctic, which are among the most quickly warming areas on Earth. Alaskan Native communities face climate-induced change, including the relocation of entire coastal villages. Elsewhere in Native America, elders tap generational memory to tell scientists how their lives have been reshaped by a rapidly warming climate.

Chapter 6, "The Inuit (and Others): If it Swims, It's Probably Poisonous begins in Nunavut, the semi-sovereign Inuit nation in northern Canada where it is snowing poison, as PCBs, dioxins, and other chemicals produced by southern industries are swept northward by prevailing winds. Southward, at Akwesasne, in northernmost New York State and nearby Canada, the fish, laced with the same chemicals, are often inedible. The Huicholes of Mexico live in a land laced with the same class of toxic pollution. In all of these locations, Native peoples have been organizing to restore a livable environment. The Inuit, in particular, have taken an active role in efforts to outlaw the use of these toxins in an international agreement: the Stockholm Convention.

Chapter 7, Alberta's Moonscape: If This Sounds Apocalyptic, It Is" describes Native peoples in Alberta, Canada, who have found some of their lands devastated by tar (or oil) sands mining. Native peoples in the United States and Canada took a leading role in opposing the Keystone XL Pipeline, which has been proposed to carry tar sands oil from Alberta to the U.S. Gulf Coast for refining and export. Blockades have been set up by the Nez Perce (in Idaho) and Lakota (at Pine Ridge, South Dakota) to block construction materials on their way to Alberta, provoking many arrests. Pine Ridge draws its water from the planned path of the pipeline. On one occasion in Idaho during 2013, the tribal chairman and most of the Nez Perce council were arrested (along with two dozen other people) as police dispersed a crowd of more than 200 people. Others, including Mik'maq bands in New Brunswick, have put their bodies on the line, with arrests, to impede fracking (hydraulic fracturing), in which high-pressure chemicals are injected deep underground to pulverize shale so that oil may be pumped out. The protests paid off after Democrat Joe Biden came into office (2020). He canceled the Keystone XL pipeline, forcing the tar sand miners to find new routes through Canada for their crudest of crude oils. As has been noted above with reference to other environmental issues, one turn of the electorate at the presidential level could open the Keystone gusher of oil, with all that implies for pollution and increases in atmospheric greenhouse gases.

Chapter 8, "Mining: Tearing at Mother's Breast" describes several of many instances in which Native peoples across North America have organized against mining on their homelands, notably in Alaska, where large salmon runs could be imperiled by the proposed Pebble Mine before a coalition of Native peoples and environmentalists obtained governmental restraint. In Wisconsin, copper, zinc, and sulfide mining are being proposed and resisted by Native peoples. Silver mining in Mexico is poisoning indigenous children. Coal strip mining is an important issue for Hopis and Navajos, as some shepherds lost many of their sheep to the toxic plumes of the Four Corners power plant (which has recently been closed, after considerable pressure from many Navajos).

Phrases such as "sacrifice areas" and "sacrifice zones" have been used to describe resource-rich Native American homelands. Winona LaDuke did the math indicating that every United States nuclear weapons test was most likely conducted on land that during the nineteenth century was guaranteed to indigenous peoples by treaty (LaDuke 1993, 99), more than 600 on Shoshone land alone (Waugh 2010). It is a supreme historical irony that native peoples whose "original instructions" have helped all of us understand the concept of earth as mother have been subject to some of North America's worst pollution caused by mining that regards "mother earth" as a "mother lode."

This book links present-day Native American cultural and economic revival to a fundamental struggle to restore the health of both Native peoples and their homelands. It links past and present with a sense of Native Americans' perceptions of nature and the sacred and, in so doing, also provides the majority society with an example to emulate as we emerge, by necessity, from the age of fossil fuels into a sustainable energy paradigm. This is an existential question of our time, and Native peoples, by necessity, are providing examples of survival. The rest of us must bear witness not only to a transition to a new model of energy generation. Also, by necessity, we all must shed the assumptions of misguided nationalism that leads not to respect of all for all, but to wars.

Awareness of economic development's costs also animates the non-Indian environmental movement. "This, then, is the nemesis that modern Western man, together with his imitators... has brought upon himself by following the directive given in the first book of *Genesis*," wrote the great English historian Arnold Toynbee. "That directive has turned out to be bad advice, and we are beginning, wisely, to recoil from it" (Toynbee 1973, n.p.).

The late scientist and public intellectual Carl Sagan said that we are unlikely to witness intelligent life from other planets not only because of distance from the many stars and planets that we now know exists. Sagan also believed, based on humankind's example, that at a certain level of technological achievement, such civilizations probably destroy themselves. Based again on our own experience, such destruction may be the result of advanced civilization's own ingenuity—for example, the machines that burn carbon and keep our lights on, but also will heat our atmosphere beyond the endurance of life. Witness, as well, the technological ingenuity that makes our radioactive weapons a tool for winning wars, but also a death sentence in which it has been said that the living will envy the dead.

So, here we are at the junction of impending visions of end time; it is, perhaps, our turn to face Sagan's paradigm. This book is a set of accounts in which Native Americans have endured death in the present tense. It is also an offering to the future generations toward whom many Native peoples direct their prayers.

Bruce E. Johansen
Omaha, Nebraska, USA
September 2022

References

Berry, Carol. "Scholar Daniel Wildcat in Discussing the Environment: It's Relatives, Not 'Resources.'" Indian Country Today Media Network, October 1, 2011. http://indiancountrytodaymedianetwork.com/gallery/photo/scholar-daniel-wildcat-in-discussing-the-environment%3A-it%E2%80%99s-relatives%2C-not-%E2%80%98resources%E2%80%99-56599. Last accessed September 22, 2014.

Grossman Zoltán and Alan Parker. Asserting Native Resilience: Pacific Rim Indigenous Nations Face the Climate Crisis. Corvallis: Oregon State University Press, 2012.

Hansen, Terri. 2013. "Major Environmental Disasters in Indian Country. Indian Country Today Media Network, October 8. http://indiancountrytodaymedianetwork.com/2013/10/08/7-major-industrial-environmental-disasters-indian-country-151661. Last accessed April 8, 2015.

Hansen, Terri. 2014. "532 Superfund Sites in Indian Country." Indian Country Today Media Network, June 17. http://indiancountrytodaymedianetwork.com/2014/06/17/532-superfund-sites-indian-country-155316. Last accessed September 4, 2018.

Hooks, Gregory and Chad L. Smith. "The Treadmill of Destruction: National Sacrifice Areas and Native Americans," The American Sociological Review. 69:4(2004):558-575.

Johansen, Bruce E. and Roberto Maestas. Wasi'chu: The Continuing Indian Wars. New York: Monthly Review Press, 1979.

Klein, Naomi. This Changes Everything: Capitalism and the Climate. New York: Simon & Schuster, 2014.

Kolbert, Elizabeth, "When 'Natural Disasters' Aren't. National Geographic, March, 2021, 15-16.

Kort, Eric A., Christian Frankenberg, Keeley R. Costigan, Rodica Lindenmaier, Manvendra K. Dubey, and Debra Wunch. "Four Corners: The Largest U.S. Methane Anomaly Viewed from Space." Geophysical Research Letters 41:19(October 16, 2014): 6,898-6,903.

LaDuke, Winona. "A Society Based on Conquest Cannot Be Sustained: Native Peoples and the Environmental Crisis." In Toxic Struggles: The Theory and Practice of Environmental Justice. Ed. Richard Hofrichter. Philadelphia: New Society Publishers, 1993, 98-106.

Lustgarten, Abraham. "Inside the Power Plant Fueling America's Drought." ProPublica in Indian Country Today Media Network, July 3, 2015. http://indiancountrytodaymedianetwork.com/2015/07/03/inside-power-plant-fueling-americas-drought-160928./. Last accessed October 11, 2017.

Minard, Anne. "Methane 'Hot Spot' Seen from Space Hovers Over Four Corners." Indian Country Today Media Network, October 28, 2014. http://indiancountrytodaymedianetwork.com/2014/10/28/methane-hot-spot-seen-space-hovers-over-four-corners-157560. Last accessed November 7, 2017.

Osborne, Tracey. "Native Americans Fighting Fossil Fuels? Scientific American, April 9, 2018. https://blogs.scientificamerican.com/voices/native-americans-fighting-fossil-fuels/. Last accessed April 6, 2020.

Regan, Shawn and Terry Anderson. "The Energy Wealth of Indian Nations." Bozeman, Montana Property and Environmental Research Center. George W. Bush Institute, 2013. https://www.

perc.org/wp-content/uploads/old/GWBI-EnergyWealthIndianNations.pdf. Last accessed June 6, 2016

Russo, Kurt. Review. Zoltán Grossman. and Alan Parker. Asserting Native Resilience: Pacific Rim Indigenous Nations Face the Climate Crisis. Corvallis: Oregon State University Press, 2012.

Toensing, Gale Courey. "Anaya Seeks Information for Study on Extractive and Energy Industries." Indian Country Today Media Network. February 6, 2013. http://indiancountrytodaymedianetwork.com/2013/02/06/anaya-seeks-information-study-extractive-and-energy-industries-147495. Last accessed May 22, 2016.

Toynbee, Arnold. "The Genesis of Pollution," New York Times, September 16, 1973, n.p.

Waugh, Charles, "'Only You Can Prevent a Forest': Agent Orange, Ecocide, and Environmental Justice." Interdisciplinary Studies of Literature and the Environment. 17:1 (2010): 113-132. http://digitalcommons.usu.edu/english_facpub/791. Last accessed October 10, 2014.

Chapter 2
Akwesasne: Land of the Toxic Turtles

Within the living memory of many people at Akwesasne, "The Land Where the Partridge Drums," has inherited the toxicological consequences of General Motors waste lagoons where animals have been found, dead and alive, with levels of PCBs in their fat that qualified them as toxic waste under U.S. Environmental Protection Agency guidelines. Akwesasne had become riskier to human health than most urban areas, a place where, if he or she knew, any partridge still living may have been more concerned about its heartbeat rather than its drumbeat (Johansen 1993, 1–3).

Until the late 1950s, The Akwesane Mohawks' homeland had been rich in natural wonders for many centuries. It was thickly forested and well-watered, home to white pine, and ash, oak and hickory, with deer, elk, and many other game animals. In the bottomlands, into rich valley soil at the confluence of several rivers allowed intensive and rewarding farming. About 1755, the Akwesasne Mohawks gave their territory the name "Akwesasne," which in the Mohawk language means "Land Where the Partridge Drums," for the distinct sound that a partridge (a.k.a. male ruffled grouse) makes during its courtship rituals. Lying at the confluence of the Saint Lawrence, Racquette, Grass, Saint Regis, and Salmon rivers, Akwesasne, until very recent times, also provided its human residents with walleye pike, sturgeon, bass, and other fish.

That was before St. Lawrence became an industrial river, and its waters were lined with producers of PCBs, dioxins, and other human inventions that were meant to contaminate and kill plants and animals, including human beings. At this point, the factories along St. Lawrence began to make a lot of money, and much of the natural environment began to die.

The St. Lawrence Seaway opened on April 25, 1959. Soon, the banks of the river became studded by a growing number of large industrial plants carrying household names such as Alcoa, Reynolds, General Motors, General Electric, and others, many of which dumped hazardous wastes at various sites near the reservation. Ignoring the pollution of the water and the land, New York officials generally supported the plants as sources of tax revenue. Little was said about chemical pollution or its effects on Mohawks and their neighbors.

Over the following years, many of the plants closed, as their owners moved away, some because their waste products, such as PCBs and dioxin, had become illegal. Many of the industrial plants' owners took what was valuable to them. Nearly always, they left ponds and swamps of toxic chemicals behind. Sooner or later Inspectors for the U.S. federal Environmental Protection Agency (EPA) and New York State moved in, finding PCBs, among other hazardous materials, and declaring Superfund sites, meanwhile testing the earth and waterways. Residents were told not to eat fish from St. Lawrence, a Mohawk staple for hundreds of years. Mothers were told not to breastfeed their babies, who were now one step up a contaminated food chain. Much of the Mohawks' homeland had become a collection of EPA Superfund sites. Native Americans were commonly becoming toxic waste site. By the time the people of Akwesasne were told that their land was one of the United States' worst toxic Superfund sites, roughly a third of such sites in the United States had reached that status. Since most of the plants were outside even by only a few yards from reservation boundaries, the Mohawks had no jurisdiction. The chemicals, however, did not respect the boundaries when they spread into the soil and underground water.

In medical circles, PCBs are known as endocrine-disrupting chemicals (EDCs), which can block (or mimic) hormones and disrupt the body's usual necessary functions. These EDCs often have a role in the formation of several cancers. These EDCs also often obstruct or otherwise interfere with people's immune, reproductive, and nervous systems. "Endocrine disruption seems to be the effect which is most far-reaching because other effects on the reproductive system may be well tied into that," said Lawrence Schell, a professor at the State University of New York at Albany (Hansen 2018).

Akwesasne, with its many toxic waste sites, became a national and international site for their study of Persistent Organic Pollutants (POPs). because of its long record of exposure and former studies. As the number of studies increases, so does the range of health maladies with which they have become associated. By 2020, researchers were studying these diseases as a group. The number of individual diseases within this group has steadily grown into its own syndrome. Zafer Aminov, and David O. Carpenter described this development in *Environmental Pollution:* "The metabolic syndrome (MetS) is a group of diseases that tend to occur together, including diabetes, hypertension, central obesity, cardiovascular disease and hyperlipidemia. Exposure to persistent organic pollutants (POPs) such as polychlorinated biphenyls (PCBs) and organochlorine pesticides (OCPs) has been associated with increased risk of development of several of the components of the MetS."

Aminov and Carpenter's study aimed to determine whether the associations with POPs are identical for each of the components and for the MetS. They examined 601 Akwesasne Mohawks between the ages of 18 and 84 who first answered a questionnaire, then were measured for height and weight, after which they provided blood samples for examination of "serum lipids and fasting glucose) and analysis of 101 PCB congeners and three OCPs [dichlorodiphenyldichloroethylene (DDE), hexachlorobenzene (HCB), and mirex]. Associations between concentrations of total PCBs and pesticides, as well as various PCB congener groups with each of the different components of the MetS were determined, to ask whether there were similar risk factors for all components of the MetS" (Aminov and Carpenter 2020). The results were not always consistent: "After adjustment for other contaminants, diabetes and hypertension were strongly associated with lower chlorinated and mono-ortho PCBs, but not other PCB groups or pesticides. Obesity was most closely associated with highly chlorinated PCBs and was negatively associated with mirex. High serum lipids were most strongly associated with higher chlorinated PCBs and PCBs with multiple ortho-substituted chlorines, as well as total pesticides, DDE and HCB. Cardiovascular disease was not closely associated with levels of any of the measured POPs. While exposure to POPs is associated with increased risk of most of the various diseases comprising the MetS, the specific contaminants associated with risk of the component diseases are not the same" (Aminov and Carpenter 2020).

Researchers also established that PCBs alter thyroid gland function at Akwesasne. Previous studies had found lower testosterone levels and established links to autoimmune disorders.

Within a sliver of a century, this land of natural wonders had become a place where one cannot eat local fish and game, because their flesh now is laced at toxic levels of PCBs and other carcinogens. In some places, one cannot drink the water, for the same reason. In parts of Akwesasne, residents have been told to plow under their gardens and to have mothers' breast milk tested for contamination. In place of sustaining rivers and a land to which the Mohawks still offer thanksgiving prayers, capitalism had offered incinerators and dumps for medical and industrial waste. Akwesasne, which straddles New York State's border with Quebec and Ontario, has become the most polluted Native reserve in Canada, and a number-one toxic site in the U.S. Environmental Protection Agency's "Superfund" list of sites requiring emergency cleanup.

The General Motors Foundry was of special interest at Akwesasne because it left behind some of the area's most noxious pollution, and dumped it very close to Mohawks' homes and fields. The G.M. plant was followed by steel mills and aluminum plants and that provided raw materials and parts for the foundry. As plants grew around the Akwesasne was named by the Mount Sanai School of Medicine as the most polluted non-military contamination site in the United States. Having studied the area, by the mid-1980s, the EPA advised adults to eat no more than half a pound of locally caught fish per week. Women who were pregnant or nursing, and young people under 15 were advised not to eat any local fish.

Soon after the St. Lawrence Seaway opened during the middle 1950s, General Motors, Reynolds Metals Company, and the Aluminum Company of America built plants directly upstream of Akwesasne. General Motors never obtained a permit to operate its dumpsites, according to the state attorney general's office. (Thomas 2001).

Paul Thompson, who has lived in Akwesasne most of his life, remembered his childhood as a cleaner and more innocent time. Near the St. Lawrence River, he watched walleye pike leap to spawn in April. His siblings bought food from fishermen on the river's edge, as they looked at fish that had been hauled into crates and picked out the evening's main course: maybe a perch, or a bass, or a sturgeon head that might go into a soup" (Sengupta 2001Nearby, Thompson's sisters and brothers foraged for scrap metal that they sold for a few extra dollars (Sengupta 2001).

At the time, no one at Akwesasne realized that a General Motors engine parts factory was turning the fish to toxic waste and the children's playground into a toxic dump. This was the 1950s, when the children nor the parents knew anything about PCBs or dioxins. Without knowing anything, the fishermen and the children had become part of a large network comprising one of the largest chemically contaminated sites for some of the deadliest poisons in the United States, because of the water's toxicity in several waste ponds that leached into the water table below much of the water table that also fed the St. Lawrence, carrying polyaromatic hydrocarbons, phenols, as well as metals and rotting organic matter to the Atlantic Ocean (Sengupta 2001).

General Motors again seemed to be the biggest contributor to the toxicity that was causing a great deal of damage to the parts of Akwesasne near the St. Lawrence, as well as anyone who ate fish from the river, because of PCBs used in manufacture of aluminum products. In 1978, however, PCBs became illegal after they were defined as a hazardous substance under the aegis of the U.S. Clean Water Act. The PCBs were also banned under the U.S. Toxic Substances Control Act because of their detrimental effects on human and environmental health, which have been associated with reproductive failure, liver damage, and skin disease.

The EPA fined General Motors for illegal use of PCBs, and declared the area as a national Superfund site. All of this sounded like concrete action that would rid the area of PCBs and other toxins, but many scientists knew better. The PCBs would not vanish from the food chain by the waving of a bureaucratic magic wand. Not to mention the fact that many federal government agencies failed to follow up with the enforcement of many regulations and cleanup requirements that had been promised. The Mohawks at Akwesasne had requested removal of all General Motors' toxic landfills within 500 yards of Mohawk land, for example, a reasonable request given the damage on the environment that they had caused. The EPA denied the Mohawks' suggestions. Instead, Instead, the landfill was capped instead; a cheaper but less effective solution.

Dissatisfaction with The EPA's lack of effective solutions prompted Mohawks on both sides of the U.S.- Canadian border form lobbying groups, of their own. One of these was the Akwesasne Task Force on the Environment, formed in 1987, keep an eye on cleanup efforts and decision-making in a joint effort with the EPA and other U.S. and Canadian governmental agencies.

Some scientists who worked within government agencies took up the cause and made the toxicity at Akwesasne a major cause in their professional lives. Ward Stone, a wildlife pathologist with the New York State Department of Environmental Conservation engaged in detailed fieldwork to answer an important question: just how toxic were animals that lived in and near the waste dumps that industries had left behind? What he found astounded nearly everyone. Some of the animals' bodies were so toxic that they surpassed the EPA's standards as toxic solid waste.

Examining animals at Akwesasne, Stone found that PCBs, insecticides, and other toxins had not been contained within designated dumps. After years of use, some of the dump sites had leaked, and the toxins had leaked into the food chain of human beings and almost every other species of animal in the area. The Mohawks' traditional economy, which was based on hunting, fishing, and agriculture, was being poisoned out of existence. Once again, the toxins were not behaving according to expectations of government agencies. They were infiltrating nearly the entire environment not only in, but around and close to the old, abandoned waste ponds. Stone, however, did not report suppositions. He waded into the swamps and ponds and tested pathological reality.

The Mohawks started Stone's toxic tour of Akwesasne at one of General Motors' waste lagoons, which had been so well-hidden from inquiry or curiosity that some maps had no name for it except "un-named tributary cove." Stone gave it the name "contaminant cove" because of the high levels\ of pollution that he found there.

At contaminant cove one day in 1985, from a Mohawk perspective, Akwesasne's environmental crisis at assumed a whole new foreboding shape. At that time, the New York State Department of Conservation caught a female snapping turtle that contained 835 parts per million of PCBs. The turtle has a fundamental significance among the Haudenosaunee (Iroquois), whose complex creation story describes the world's creation on a turtle's back. To this day, many Iroquois call North America "Turtle Island." This usage has spread to many other Native peoples on Turtle Island. The fact that attempts to clean the environment of PCBs, et al. had been unsuccessful carried a new foreboding shape for the Mohawks, and everyone else who calls the world "Turtle Island" after that.

Working teams had been unable to clear much of Akwesasne's PCB contamination, as studies piled up describing how the toxicity was changing the lives of the people who are afflicted with this poison—what they eat, and how it may affect them long term. Medical and health professionals have described again and over how dangerous PCB exposure continues to be, building knowing anxiety among the people. Studies revealed that the bodies of Akwesasne's adult population had about three times the background levels of PCBs compared to the general United States population. The comparison was not really significant, however, because the general level was nearly nonexistent.

The situation at Akwesasne and its intensity drew a large number of health reports by experts in several academic fields. The reports were good for describing the dire conditions of how acute the situation had become but did little to clear the poisons from peoples' bodies.

A few studies (usually newspaper stories about EPA reports) indicated that the deeper workers dug, the more contamination they found. The reports were thick with medical jargon and of little use to non-professional people.

Akwesasne mothers' milk was examined several times and found to be too toxic for use—ominous for mothers who lost their richest source of nutrition for their babies. Canned formula was very expensive and very difficult to find in areas with very few commercial grocery stores. The same problem was encountered by as adults absorbed recommendations that they stop or severely curtail consumption of their dietary staples. Quitting consumption of whale and seals was like telling lower-latitude people to quit eating beef, chicken, turkey, and most other sources of protein.

An early study that was published during 1998 in the *American Journal of Epidemiology*. This study was undertaken "to determine the relation between the consumption of contaminated local fish and concentrations of total polychlorinated biphenyls (PCBs) and 68 PCB congeners in the milk of nursing Mohawk women residing near three hazardous waste sites." Contamination was relatively high to begin, but decreased as warnings about PCB's probable effects were heeded. (Fitzgerald et al. 1998.)

Articles on the same theme appeared during the next quarter century to include women with similar contamination issues.

- Relationship of lead, mercury, mirex, dichlorodiphenyldichloroethylene, hexachlorobenzene, and polychlorinated biphenyls to timing of menarche among Akwesasne Mohawk girls. Denham M, Schell LM, Deane G, Gallo MV, Ravenscroft J, DeCaprio AP; Akwesasne Task Force on the Environment. Pediatrics. 2005 Feb;115(2):e127–34. doi: 10.1542/peds.2004-1161. Epub 2005 Jan 14. PMID: 15653789
- Environmental and occupational exposures and serum PCB concentrations and patterns among Mohawk men at Akwesasne. Fitzgerald, EF, Hwang SA, Gomez M, Bush B, Yang BZ, Tarbell A.

    J Expo Sci Environ Epidemiol. 2007 May;17(3):269–78. doi: 10.1038/sj.jes.7500500. Epub 2006 May 31.

    PMID: 16736058.
- Polychlorinated biphenyl (PCB) exposure assessment by multivariate statistical analysis of serum congener profiles in an adult Native American population. DeCaprio AP, Johnson GW, Tarbell AM, Carpenter DO, Chiarenzelli JR, Morse GS, Santiago-Rivera AL, Schymura MJ; Akwesasne Task Force on the Environment.

    Environ Res. 2005 Jul;98(3):284–302. doi: 10.1016j.envres.2004.09.004. PMID: 15910784.
- Effects of polychlorinated biphenyls on the nervous system. Faroon O, Jones D, de Rosa C. Toxicol Ind Health. 2000 Sep;16(7–8):305–33. doi: 10.1177/074823370001600708. PMID: 11693948 Review.
- Epidemiologic studies of PCB congener profiles in North American fish consuming populations. Chiu A, Beaubier J, Chiu J, Chan L, Gerstenberger S. J Environ Sci Health C Environ Carcinog Ecotoxicol Rev. 2004 May;22(1): 13–36. doi: 10.1081/GNC-120038004. PMID: 15845220 Review. Last accessed March 20, 2022. https://pubmed.ncbi.nlm.nih.gov/30359956/
- Relationship of thyroid hormone levels to levels of polychlorinated biphenyls, lead, p,p'- DDE, and other toxicants in Akwesasne Mohawk youth. Schell LM, Gallo MV, Denham M, Ravenscroft J, DeCaprio AP, Carpenter DO. Environ Health Perspect. 2008 Jun;116(6):806–13. doi: 10.1289/ehp.10490.PMID: 18560538

- McRae, Kim Ellen, "Effects of PCB Contamination on the Environment and the Cultural Integrity of the St. Regis Mohawk Tribe in the Mohawk Nation of Akwesasne" (2015). *Graduate College Dissertations and Theses.* 522. https://scholarworks.uvm.edu/graddis/522 Last accessed March 25, 2022.
- Ravenscroft J, Schell LM; Akwesasne Task Force on the Environment. "Patterns of PCB Exposure Among Akwesasne Adolescents: The Role of Dietary and Inhalation Pathways." REnviron Int. 2018 Dec;121(Pt 1):963–972. doi: 10.1016/j.envint.2018.05.005. Epub 2018 October 22. PMID: 30359956.Last Accessed May 1, 2022.

Ravenscroft et al. found that "Even relatively low levels of fish consumption within the composite dietary matrix of adolescents at Akwesasne remains a pathway of exposure to postnatally acquired PCBs. In addition, there is evidence of an unidentified, perhaps airborne, exposure pathway that warrants further attention as this congener profile accounted for 50% of the total variance within the adolescents' serum PCB levels" (Ravenscroft and Schell 2018).

Intensive pollution of Akwesasne became evident through many other studies. Publicity helped to make the issue well known, but did little to clean up the lethal mess that General Motors had left behind. In 1995 the Mohawk Adolescent Wellbeing Study, known as MAWBS, with the Akwesasne Mohawk Nation, investigated human health effects of PCBs,) with two New York State Superfund Sites (Reynolds Metal Company and Aluminum Company of America (ALCOA) This study was conducted at an important site upstream of Mohawk territory, at the St. Lawrence River's juncture of Quebec, Ontario, and New York State. Rates of compliance with fish advisories during the 1980s that caused disruption of the Mohawks' traditional lifestyle—useful information, for sure, but having very little direct influence on actually cleaning up the pollution itself.

Previous studies have found memory deficits in children following with PCB exposure. Studies of children exposed to Lake Michigan fish found memory deficits (of different types) during repeated testing from infancy (Jacobson et al. 1985), in preschool (Jacobson et al. 1990), and pre-adolescence (Jacobson and Jacobson 1996). Children eating fish from Lake Ontario had memory deficits in memory only during infancy (Darvill et al. 2000) and not as preschoolers (Stewart et al. 2003. Cognitive deficits that were found in these studies were attributed to prenatal PCB exposure. Once again, this information is valuable for medical, but children are still being exposed, running a risk of severe deformations.

Back on the ground at Akwesasne, during 1985, Ward Stone, cooperating, with the Mohawks, found a masked shrew that somehow had managed to survive with a P.C.B. level of 11,522 parts per million in its body, the highest level that Stone had detected in a living creature, 250 times the minimum standard to qualify as hazardous waste. Stone and the Mohawks to established Akwesasne as the major PCB-impacted zone in the United States. No one gives trophies for such a distinction, of course. It *did* ruin the area's long-standing reputation as a good place to catch tonight's dinner.

Before news of Akwesasne's pollution had become well known, Akwesasne had been home to more than 100 commercial fishermen, and about 120 farmers. By 1990, fewer than 10 commercial fishermen and 20 farmers remained.

The Environmental Protection Agency released its Superfund cleanup plan for the General Motors foundry during March 1990. The cost of cleanup was estimated at $138 million, making the General Motors dumps near Akwesasne the costliest Superfund cleanup job in the United States, number-one on the E.P.A.'s list as the United States' worst toxic dump. By 1991, the cost was reduced to $78 million. Even so, the General Motors dumps were still ranked as the most expensive toxic cleanup under supervision by the EPA.

While no federal standards exist for P.C.B.s in turtles, the federal standard for edible poultry is 3 per million, or about one-third of 1% the concentration in that snapping turtle. The federal standard for edible fish is 2 parts per million. In soil, on a dry-weight basis, 50 parts per million is considered hazardous waste, so that turtle contained roughly 15 times the concentration of PCBs necessary, by federal standards, to qualify its body as toxic waste. During the fall of 1987, at about the same time, Stone found another snapping turtle, a male, containing 3067 parts per million—1000 times the level advised in domestic chicken, and about 60 times standard for hazardous waste. Contamination was lower in female turtles because they shed some of their own contamination by laying eggs, while the males stored more of what they ingested as body fat.

"We can't try to meet the challenges with the meager resources we have," said Henry Lickers, a Seneca who was employed by the Mohawk Council at Akwesasne. Lickers became a mentor to younger environmentalists at Akwesasne. He also became a leader in the fight against fluoride emissions from the Reynolds plant. "The next ten years will be a cleanup time for us, even without the money," said Lickers (Johansen 1993, 19). Destruction of Akwesasne's environment was credited by Lickers as being the catalyst that spawned the Mohawks' deadly battle over high stakes gambling and smuggling. "A desperation sets in when year after year you see the decimation of the philosophical center of your society," he said (Johansen 1993, 19).

Even with all the talk, all the medical journal articles, all the EPA discussions, and all the digging, PCB contamination continues to be a persistent problem at Akwesasne. In 2014, the EPA said that PCB pollution could be 250% higher than earlier estimates, as people dig more test holes, and find that the chemical has leached both downward and outward from old waste dumps. Akwesasne have continued to edge higher. By 2000, the EPA had three Superfund sites, not one. As the EPA continue to look, the scope of the disaster grew. In the meantime, expenses to search for more PCBs were outrunning EPA's budget, in many cases long before actual cleanup had been started. The sheer volume of PCB pollution was exceeding the EPA's ability to account for it, much less clean it up.

For example, federal toxicity regulations for PCB contamination began at 10 parts per million, but some of the 335,000 tons of untreated sludge excavated at 40 feet down was coming in 500,000 parts per million, close to pure PCB.

The Mohawks were not alone. As cities and towns all other the United States increasingly use restrictive environmental regulations enacted by states and cities were bringing polluters to native reservations. "Indian tribes across America are grappling with some of the worst of its pollution: uranium tailings, chemical lagoons and illegal dumps. Nowhere has it been more troublesome than at...Akwesasne," wrote Rupert Tomsho, a reporter for the *Wall Street Journal*. At Akwesasne, Katsi Cook, a Mohawk midwife who has studied the degree to which mothers' breastmilk has been laced with P.C.B.s at Akwesasne said "This means that there may be potential exposure to our future generations. Analysis of Akwesasne Mohawk mothers' milk shows that our bodies are, in effect, part of the [General Motors] landfill".

Nearby, Turtle Cove, an inlet leading into the St. Lawrence River, was a favorite swimming hole for children at Akwesasne. In the spring, boys, like generations of men before them, learned to spear bullhead pike making their way through the cove to spawn. The cove, only a few feet from the General Motors foundry, was a swimming hole no longer. Instead, it was one of General Motors' toxic-waste dumps.

Akwesasne resident Dana Leigh Thompson grew up with a 40-foot General Motors waste heap as a neighbor. The toxic hill slopes into "Containment Cove," which was a local swimming hole until tests revealed PCB levels many times toxic limits. "There were three big rocks out there," Thompson said. "When we taught kids how to swim, they could swim out to the middle and stand. It was an achievement" (Seely 2001, n.p.).

Thompson and other local residents began to suspect toxicity in G.M.'s waste dumps during the middle 1970s, but General Motors continued to dump PCBs in the area without a state permit until 1986. Cleanup efforts began about 1988, but have stalled over differing approaches to the problem. During 1988, according to residents of Akwesasne, "A crew of men, covered head to toe in white spaceman-like suits, covered it [the mound] with an impermeable sheath" (Sengupta 2001). Meant to be temporary (in place until the dump was cleaned up) the capped mound remained in place 13 years later. The installation of the "temporary" cap, during 1983, initiated Thompson's former playground into the ranks of federal Superfund sites, as one of the most toxic (in this case, PCB-laced) patches of ground in North America.

At first, "I didn't even know what PCBs were," said Jim Ransom, an Akwesasne resident and director of the Haudenosaunee Environmental Task Force, a group that advocates for all Haudenosaunee (Iroquois). "There was a high level of concern, but I think that there was also a lot of unknowns because people didn't know what this chemical was and what it could do to us." By the mid-1980s, preliminary testing showed that it was no longer safe to eat fish and wildlife caught in some areas of the reservation. Sheree Bonaparte, then a young mother with a farm near the G.M. landfill, laughed when G.M. first distributed bottled water to residents. At first, she said, "Everybody kind of thought it was ridiculous. The water comes from the earth and it seemed silly to go get it from a bottle" (Thomas 2001).

The Mohawks, state agencies and area universities soon began studying PCB levels in mothers' breast milk and in infants. "Those studies proved beyond any shadow of a doubt that at the beginning of the study, the Mohawks had significantly higher levels of PCBs," said David Carpenter, a professor of environmental health and toxicology at SUNY Albany. (Thomas 2001). Carpenter said that the Akwesasne Mohawks have higher-than-average rates of some diseases that are associated with PCB contamination. One such disease, hypothyroidism, is "strikingly elevated," Carpenter said. (Thomas 2001). The following symptoms of hypothyroidism develop when the thyroid gland produces too much T3 and T4 hormone, driving up metabolism and causing a variety of symptoms of an over-active thyroid that can lead to cancer. This disease also sometimes leads to learning disabilities, including mental dullness, and obesity in children. PCBs disrupt the production of thyroid hormones, which leads to hypothyroidism.

For more than a decade, mainly during the mid-1980s into the 1990s, General Motors and the Mohawks debated how to best clean up the waste lagoons. The company suggested sealing the dumps permanently in place, meanwhile also building a wall to prevent existing PCBs from migrating to other parts of Akwesasne. Federal officials approved this plan, but General Motors required access to the reservation to build the wall. The Mohawks denied access because they believed that G.M. was seeking a relatively inexpensive way out of a problem that required removal, *en masse*, of all soil tainted by PCBs dumped there.

In the meantime, the EPA commended General Motors for moving diligently to clean up its waste sites. General Motors found itself inching toward agreement with the Mohawks' solution. The company, for example, in 1995 did remove about 23,000 cubic yards of polluted sediment from the St. Lawrence River during 1995. During the year 2000, General Motors excavated contaminated sludge from inactive lagoons.

"This is the only place we have, and we're going to be here forever," explained Ken Jock, director of the St. Regis Mohawk tribe's environmental division. "Our teachers have told us, when we make a decision we have to look at how it affects the next seven generations. It's a different sense of time" (Sengupta 2001). Before the area was so widely contaminated, fishing, hunting, and trapping "were something our parents had pride in handing down to our children," said Jock. "Because Akwesasne residents have been advised by state officials not to eat local fish or game, fishing and hunting skills are being lost. "It's pretty important to our identity as a people" (Thomas 2001).

Examining several Akwesasne residents, who had diseases that were believed to be caused at least in part by PCBs, scientists concluded that even low levels of PCB exposure could cause more serious illnesses than they had previously thought. "That small relationship we expect to see correlated with reduced I.Q., with poor performance in school, with some abnormality in growth, particularly sexual maturation, and increased susceptibility to certain chronic diseases such as thyroid disease and diabetes," said Carpenter. "This has adversely affected their health" (Sengupta 2001).

The New York Times reported that Thompson's family had been continually wracked by several illnesses that had long been very rare at Akwesasne. Thompson himself had diabetes. Four of his five siblings had thyroid disorders of a type often aggravated by PCBs. Thompson's sister Marilyn had her thyroid gland removed when a tumor was discovered there. All six of her children had asthma; two of them also had learning disabilities; another suffered from the thyroid condition described above. A two-year-old granddaughter of Thompson was born with a muscle disorder that affected her motor skills. Another family member had experienced 14 miscarriages (Sengupta 2001).

Rowena General of Akwesasne said the contamination led to a "health crisis" for the more than 10,000 people who live on the reservation. "Recent analysis of clinical and hospital records on the reservation indicated an epidemic of thyroid problems and also the incidence of cancers, diabetes and respiratory diseases is higher than average," General said (Associated Press 2001, March 5).

Distribution of chronic diseases at Akwesasne was studied using computerized medical records of the St. Regis Mohawk Health Services Clinic. Annual and five-year incidence rates were computed for the period January 1, 1992, to January 1, 1997, for asthma, diabetes mellitus type II, hypothyroidism, and osteoarthritis. The study indicated that hypothyroidism and diabetes "showed higher age-specific prevalence than in the general U.S. population. Osteoarthritis was extremely frequent among people 60 years of age and older, and it may also be elevated in prevalence in relation to the U.S. general population. The incidence and prevalence trends of diabetes type II and osteoarthritis were stationary, but those for asthma and hypothyroidism showed increases over the study period. Morbidity from asthma and acquired hypothyroidism should be monitored in the future and investigated through analytic epidemiologic methods for a possible association with lifestyle and environmental factors" (Negoita et al. 2001, 84).

A study at Cornell University added that smokestack effluvia from the Massena Reynolds Metals factory had destroyed cattle and dairy farms in Cornwall on the Ontario side of Akwesasne that had been profitable. The same study associated fluorides with the deaths of once-healthy of cattle as early as 1978. Many of the cattle, as well as fish, had died from fluoride poisoning that weakened their bones and decayed their teeth. Ernest Benedict's Herefords died while giving birth, while Noah Point's cattle lost their teeth and Mohawk fishermen landed perch and bass with deformed spines and large ulcers on their skins. The fluoride was a by-product of a large aluminum smelter in Massena, New York, that routinely fills the air over Akwesasne with yellowish-gray fumes that smell of acid and metal (Krook and Maylin 1979, 1).

Even though the plant cut its average fluoride emissions from about 200 to 300 pounds per hour in 1959 to 75 pounds in 1980, the ever-smaller cattle population still feeding up the hill from the plant continued to die of fluoride poisoning. The health populations of Akwesasne (including cattle), have long been aggravated by the geographic position of the plants that emit toxins, which are located west of the reservation, upstream, and often upwind from people, animals, and plants. Fluoride poisoning of cattle on Cornwall Island cattle was "manifested clinically by stunted growth and dental fluorosis to a degree of severe interference with drinking and mastication. Cows died at or were slaughtered after the third pregnancy....

Concentrations exceeding 10,000 p. p. m. fluoride were recorded in cancellous bone of a 4-to 5-year-old cow" (Krook and Maylin 1979, 1).

Late in March 2001, New York State Attorney General Eliot Spitzer and the St. Regis [Akwesasne] Mohawk Nation served a legal notice to General Motors that it would sue them in federal court unless cleanup of two PCB dumpsites at its Massena plant had begun made substantial progress within 90 days. In a letter to G. M., Spitzer said if the company did not make substantial progress within 90 days, he would ask a U.S. District Court judge in Albany to declare the site an "imminent and substantial endangerment" and order an immediate cleanup (Associated Press 2001, March 5). "General Motors has been on notice since at least 1980 that PCBs were being released into the St. Lawrence River and onto the St. Regis Mohawk Reservation from its two hazardous waste dumps," Spitzer said. "The company also has known for the past 15 years that the landfills may endanger public health and the environment. Despite this knowledge, General Motors has failed to control the release of these toxins from its property," he said (Associated Press 2001, March 5).

By June, barely within its 90-day deadline, the Attorney General's office said that General Motors was taking PCB-removal talks with a new sense of seriousness. "If we have to, we are ready to file a lawsuit at a moment's notice, and G.M. knows that we are prepared to do so, if necessary," said Marc Violette of the New York Attorney General's office (Associated Press 2001, June 11).

Chris Amato, an assistant New York attorney general said that the injustice is clear. "This is another example of a Native American community being treated as second-class citizens," Amato wrote. He said, additionally, that, "I guarantee you if this site was located next to a very middle-class, white neighborhood, this site would be well on its way to being remediated" (Thomas 2001).

"General Motors' illegal industrial waste dump has been poisoning the Mohawk people for over 50 years," said Akwesasne Mohawk Loran Thompson. "Despite all of our efforts, the G.M. facility continues to discharge toxic contaminants into the Akwesasne environment. General Motors is guilty of environmental injustice, and they have been completely negligent in overlooking the damages to the health, well-being, economy, and lifestyle of the Mohawk people," he said (Associated Press 2001, March 5).

A reader may note that most of the references in this section are about 20 years (that is to say, a generation) old. Mr. Thompson notes at the paragraph's beginning that "General Motors' illegal industrial waste dump has been poisoning the Mohawk people for over 50 years." That is 70 years, the life span of a lucky man or woman at Akwesasne, who had evaded most of the area's environmental atrocities. Attempts have been made to solve the problems inherent in the G.M. waste dump, but the problem has not been solved, so please excuse some Mohawks' acute cynicism when the subject comes up. A lot of well-meaning white people have come and gone at Akwesasne, but it continues to be a poster child for some of the most noxious pollution on the planet.

By 2012, Ward Stone, an environmental official for New York State, who had first detected near-lethal levels of PCBs at Akwesasne in the 1980s, had suffered several strokes, as the Mohawks at Akwesasne paid tribute to his research. Tom

Sakokwenionkwas Porter shared Akwesasne peoples' appreciation of Stone and his work: "We would like to take this opportunity to express our utmost gratitude to Ward Stone for all that he has done to help the Mohawk people. He has been a strong advocate for the health of Mohawks, especially at Akwesasne". In the 1980s, a midwife from Akwesasne...[Tekatsitsiakwa] Katsi Cook got in touch with Ward Stone and expressed concern about the effects of industrial pollution on the health of the people who reside there. Stone found extremely high levels of PCBs, insecticides and other toxins in area fish and wildlife. His work led to irrefutable proof that the dumping of contaminants by nearby factories was responsible for the high level of PCBs found in mother's milk at Akwesasne. As a result, the people at Akwesasne are benefitting from the awareness of what needs to be done to maintain a healthy environment for the generations to come. Ward Stone has worked tirelessly not only as a strong and dedicated spokesperson for the animals, insects, fish, birds, water, air and Mother Earth, but he has also spoken up for the health and welfare of the people of the Northeast. *Niawenko:wa (*thank you), Ward Stone (Mohawks Express 2012).

Decades after PCBs were banned, people at Akwesasne still carry them in their bodies at levels double the national average. "Significantly higher levels of PCBs were found among individuals who were breastfed as infants, were first born, or had consumed local fish within the past year," the study found. The study "also revealed significantly higher levels in those who had eaten fish within the previous year, (2011) in those who were first-born, and in those who were breast-fed" (Hansen 2011). "What this study is saying is that these chemicals are extremely persistent in people," said co-author Lawrence Schell, a professor at the State University of New York (SUNY at Albany). "Once you're exposed it's difficult to remove that exposure burden" (Hansen 2011).

The Grasse River Alcoa site at Akwesasne continued to affect Mohawk life. "Though the river has been awaiting cleanup for 20-plus years," reported the Indian Country Today Media Network on January 21, 2013, "The only remediation recorded on the EPA website has been the 1995 removal of about 8,000 pounds of P.C.B.s from the facility" (Mohawks Say 2013). Various cleanup proposals range from simply letting nature take its course, to a $1 billion cleanup to be billed to Alcoa. The U.S. Environmental Protection Agency in 2013 had opted for a $243 million plan to dredge some of the contaminated areas and cap others. The St. Regis Mohawk (Akwesasne) tribe favors a more expensive option that would dredge 7.2 miles of the Grasse River five feet deep. "Capping is not a permanent remedy, and ice scour is a constant threat to any cap in the Grasse River," Akwesasne Environment Division Director Ken Jock said. "Therefore, we do not support the capping of the highly contaminated sediments in main channel. Nobody has any real-world evidence that a cap can withstand a major ice jam and ice scour" (Mohawk Government 2012).

Larry Thompson, 56 in 2011, tired of waiting for General Motors to clean up a Superfund site next to his family's home at Akwesasne, was arrested in 2011 for taking a backhoe into a toxic landfill on August 11 of that year. He was charged with several misdemeanors and felonies (second-degree criminal mischief, resisting arrest and reckless endangerment) "after he drove onto the notoriously polluted mound, scooped up contaminated soil and loaded it into railroad cars that were waiting to cart away debris from the GM building that is being torn down in the wake of bankruptcy proceedings" (Mohawk Man Arrested 2011). Thompson's wife, Dana Leigh Thompson, told Indian Country Today Media Network: "She has since received numerous calls from people offering to donate time, backhoes and excavators to continue the protest." "Larry was given this order by the clan mother. Bear Clan Mother," Dana Thompson said (Mohawk Man Arrested 2011).

After all of this digging, politicking, and pledging to clean up, the landfill, formally known as the General Motors–Central Foundry Division Superfund Site at Massena, New York, *still* was one of the country's most severely contaminated toxic sites. It had been capped with plastic, clay, and soil, and planted with grass and trees. Dana Thompson said "it looks like any bucolic scene in rural New York State. But underneath lies a pile of chemicals and PCBs left over from the plant's heyday" (Mohawk Man Arrested 2011). "For 32 years we've been waiting for them to clean it up," Dana Thompson said. "Remove the pile, remove the Superfund site, take all their poison out of here and put it into a secure site," Thompson said. "General Motors' illegal industrial waste dump has been poisoning the Mohawk people for over 50 years, because they're just covering it up" (Mohawk Man Arrested 2011).

Elizabeth Hoover brought the situation at Akwesasne up to date in *The River is In Us: Fighting Toxics in a Mohawk Community* (2017), finding that while some of the pollution has been halted, the persistent nature of PCBs and other pollutants is making true cleanup very difficult. Many people still avoid tumor-ridden fish (traditional staples such as sturgeon, perch, walleye, and bullhead) because of contamination. Hoover's study is comprehensive, historically sound, richly detailed, and sensitive to the Mohawks' perception of the crisis as they wage a continuing struggle to rehabilitate their homeland, rallying to face an existential threat. Doug George Kanentiio, editor, activist, and long-time resident of Akwesasne, confirmed as much during an email to the author during April of 2018: "In the past two weeks there have been over a dozen deaths at Akwesasne, many of which are cancers and other illnesses related to the pollution there. Jake Swamp's son Andy, 55, was one of them as was my friend Ronnie Lazore, 64 and many others. Cancers, bone diseases, diabetes—all connected and we are dying in disturbing numbers. I believe Akwesasne should be abandoned and the people relocated," said George.

One compelling source is noted PCB contaminant researcher David O. Carpenter, M.D., a neurotoxicologist and professor in the Department of Environmental Health and Toxicology in the School of Public Health at the State University of New York. Carpenter studied PCB pollution on and near Akwesasne for many years. He noted that in 2013, the International Agency for Research on Cancer (IARC) reached a consensus for classifying PCBs as carcinogenic to humans. This is the highest ranking for carcinogens that the IARC uses. Additionally, he explained the World Health Organization (WHO) now considers oral and respiratory exposure to PCBs to be major exposure routes to both human beings and wildlife living near contamination sites. All of this means that Akwesasne residents are forced to live near very dangerous waste substances.

The SUNY researcher maintains landfills like this one are an inherent health threat. "PCBs volatize and escape into the air. The landfill should be moved totally away from the reservation," Carpenter said (Kader 2014).

James Ransom, a former member of the St. Regis Tribal Council said the issue involved in the cleanup is money. "The real crime that's been committed here is by General Motors. The United States basically let them shed their toxic assets. Communities like Akwesasne pay the price for that decision," Chief Ransom explained (Kader 2014).

Kakwerais believes that federal prosecutors would go after the corporations guilty of these environmental crimes. Instead, these agencies would rather prosecute the victims to remain living on contaminated land instead of helping them. "It's the legacy of colonialism's divide-and-conquer tactics," she concludes (Kader 2014).

A Ph.D. dissertation by Kim Ellen McRae of the University of Vermont (2015, in Natural Resources) describes the damage of PCBs to Akwesasne culture as well health, as well as a lack of attention to this situation in much of the environmental justice movement, It concluded: "Strong parallels can be drawn as a result of an analysis of environmental justice literature, since native communities have not, traditionally, been included in the scholarly academic literature on the Environmental Justice Movement in the United States. In addition to information gathered from institutional policy actors and related stakeholders, in-depth interviews with community members revealed a community framework for future policy development and action. Finally, the research focuses on how those community voices articulate the impacts of PCB contamination on the natural resources in the area, and as a result, on the ability of the St. Regis Mohawk tribe to maintain their culture, heritage, ceremonies, and traditional way of life" (McRae 2015).

As persistent organic pollutants, PCBs are astoundingly difficult to wrest from the bodies of human beings and other animals. Evidence of this difficulty was provided by a study published in 2018 by a medical team that examined adolescents there who were born well after cleanup efforts began. This evidence supports other findings among U.S. veterans of the Vietnam war as well as indigenous peoples there (as well as their children and grandchildren) who were copiously exposed to PCB-laden Agent Orange (also a PCB compound), described in Chapter Two of this work. The team which examined the Akwesasne concluded: "Even relatively low levels of fish consumption within the composite dietary matrix of adolescents at Akwesasne remains a pathway of exposure to postnatally acquired PCBs. In addition, there is evidence of an unidentified, perhaps airborne, exposure pathway that warrants further attention as this congener profile accounted for 50% of the total variance within the adolescents' serum PCB levels" Ravenscroft and Schell (2018).

In 2014, the EPA presented a "progress report" on results of cleanup efforts at the waste site and learned that levels of contamination in some contaminated areas were as much as 250% *above* prior reports. Because the volume of PCBs excavated was much more than anticipated, the project also was considerably behind schedule and over budget. Anne Kelly, of the EPA Emergency and Remedial Project Response Division, told Mohawks at Akwesasne. The numbers were staggering. As described in the indigenous newspaper *Two Row Time*s:

For 32 years we've been waiting for them to clean it up," stated Kakwerais, who is also known as Dana Leigh Thompson. "Remove the pile, remove the Superfund site, take all of their poison out of here and put it into a secure site," she said. "They call it a cleanup but it's really a cover-up", Kakwerais noted (Kader 2014).

"It's called environmental genocide. We can't wait longer because our people are dying. Our children are being born without their minds. All this poison goes to our land," Kakwerais (who is also known as Dana Leigh Thompson, said (Kader 2014).

"Genocide" is a strong word, but the evidence here supports it. Studies on recorded levels of testosterone in Akwesasne men and adolescent males indicate that PCBs in Native foods lead to decreases in testosterone levels in both groups. Most definitions of genocide require a motive. The fact that a people are dying is not enough for the specialists. Even so, pollution can be deadly, and death is dead, even if we have to dig for a technically acceptable motive.

While this news report cited above was 8 years old as of this writing, a check of the literature indicates that by 2022 the EPA was still digging and, with the best of bureaucratic intentions, finding that General Motors had been a very prolific generator of chemicals that will make this part of Mohawk Country very risky for human habitation for many years.

It has become abundantly clear that PCB exposure is a life sentence to victims of several cancers and other maladies which lasts a lifetime, for several generations.

Toxic contamination of traditional foods has become an issue in Alaska, as well as in Nunavut, Akwesasne, and other Native homelands. Some Alaska natives are avoiding their traditional foods out of fear that wild fish and game species are contaminated with pesticides, heavy metals and other toxins, native delegates told an international conference on Arctic pollution in Anchorage May 1, 2000.

"I have a son who has quit eating seal meat altogether," said St. Paul Island resident Mike Zacharof, president of the Aleut International Association. The association was formed because indigenous people in Western Alaska and Russia are worried about pollution that crosses their countries' boundaries, Zacharof said. According to a report in the Anchorage *Daily News*. Zacharof said: "Fear that seal livers may contain mercury has made many islanders wary of eating the staple of their diet, though most still do" (Dobbyn 2000).

"Many people in Prince William Sound no longer eat their traditional foods because of the {Exxon Valdez] oil spill. This impacts not only our physical well-being but our emotional and spiritual lives as well," said Patricia Cochran, director of the Alaska Native Science Commission (Dobbyn 2000). According to a report in the Anchorage *Daily News*, Cochran said that "Natives from every region of Alaska have been noticing more tumors, lesions, spots and sores on land and sea animals" (Dobbyn 2000). Indigenous reports are being compiled in a report for the Alaska Native Science Commission, called the Traditional Knowledge and Contaminant Project, started in 1997. The report was funded by the U.S. Environmental Protection Agency.

Research by the Circumpolar Arctic Monitoring and Assessment Program has described higher than normal levels of D.D.T. in the breast milk of Russian mothers in Arctic regions. Another study, of pregnant Alaska Native women who eat subsistence foods, conducted by researchers at the University of Alaska at Anchorage, indicates that their fetuses may be exposed to potentially dangerous pollutants. "There's no question that people are concerned not only about what it's going to do to them but to their unborn children," Cochran said (Dobbyn 2000). The Tanana ChiefsConference, a tribal social services agency for the Interior [of Alaska], has found unusually high levels of D.D.T. in salmon.

Because of toxic contamination, Alaskan Inupiat hunters now examine animals more closely than they prepare to butcher."It is called 'playing doctor,'" wrote David Hulen, reporting for the *Los Angeles Times,* describing one hunter as he "slips on a pair of rubber gloves, wipes clean a titanium-blade knife and begins cutting tissue samples from the seal to be sent off and tested for P.C.B.s, D.D.T. and nearly 50 other industrial and agricultural pollutants" (Hulen 1994, A-5). The hunters, in this case, were accompanied by Paul Becker, a Maryland-based scientist from the National Marine Fisheries Service. Welcome to the brave new world of marine mammal life guarding in the Arctic, less than a century before human beings invented environment atrocities such as PCBs.

Becker has been working with native hunters since 1987, collecting samples for a national marine mammal tissue bank. He has trained several groups of villagers in Nome, Barrow and other settlements to take samples as part of their regular hunts. Hulen reported that "All of this has caused no small amount of anxiety in the villages, where oil from rendered seal fat is a staple consumed like salad dressing and where people routinely dine on dishes such as dried seal meat, walrus heart and whale steaks" (Hulen 1994, A-5).

Charlie Johnson, executive director of the Eskimo Walrus Commission, a native group that works with the government to help manage Pacific walrus populations, said: "We've seen a huge increase in cancer rates. People hear about these things turning up and wonder if that has anything to do with it. I don't think there's enough known yet to say there's a problem here in Alaska...but I think we have to be aware that we could have a problem eventually" (Hulen 1994, A-5). Because of their role in handling body wastes, the kidneys of mammals are especially vulnerable to accumulation of toxic chemicals. Natives continue to eat the kidneys of the bowhead whale, however, because the kidney is one of the best-tasting parts of the animal.

Inuit elders describe the coming of *qallunaat,* the non-Inuit money economy in natural metaphors, as a time when the old world shifted, "They say, like a storm they could not read in the clouds....Against the wind, the people first had to brace themselves, then they had to adapt, then they had to try to stop the wind." The change has come with a stunning swiftness, in some cases within a generation or two. According to one report,

> Jamaise Mike was born in 1928 in an igloo, but now he is sitting in a Kentucky Fried Chicken restaurant in Pangnirtung, Nunavut, population 1300, on Baffin Island about 100 miles north of Iqaluit. "Most elders were born and raised on the land," Mike said, through an interpreter, speaking in the Inuktitut language. "Everything people needed to survive surrounded them on the land. Today it is different, living in a community with a store. Everything is different from when you had to do everything yourself to survive. You depended on yourself. Now, you need money" (Brown 2001, A-1).

As in many other parts of the world, the story of indigenous environmentalism is shaped, first and foremost, by the collision of the money economy and traditional, environmentally based, economies. Many Inuit are the first generation off the land, having moved to towns at the behest of the Canadian federal government. In some cases, the government's agents "systematically killed many of the dogs that pulled their sleds, giving people no choice but to come in from the land. Children were put in Christian boarding schools." "White people from the south," said Mike, "were more terrifying than polar bears" (Brown 2001, A-1).

The collision of cultures can be tragic. Where, a few decades ago, consumption of alcohol was rare, Baffin Island Inuit now have Canada's highest rates of drunkenness, as well as suicide. In 1999, 58 of Nunavut's 27,000 people committed suicide. Fifty-two were by hanging, six by firearm. Fifty-seven of them were Inuit (Brown 2001, A-1).

As the economic system of the "south" sets down roots, the Inuit sense a more acute need to preserve traditions that are slipping away. "We live in wooden houses, drive Jeep Cherokees, and fly in jumbo jets all over the world. But we are still Inuit. It is our spirit, our inner being, that makes us Inuit," said John Amagoalik, former chief commissioner of the Nunavut Implementation Commission (Brown 2001, A-1).

References

Aminov, Zafer and David O. Carpenter. "Serum Concentrations of Persistent Organic Pollutants and the Metabolic Syndrome in Akwesasne Mohawks, a Native American Community." *Environmental Pollution* vol. 260 (2020) p. 114004. Last accessed May 16, 2022.

Associated Press in Syracuse.com [Syracuse newspapers]. "State, G.M. Talking so Lawsuit is Set Aside". June 11, 2001. http://syracuse.com/newsflash/index.ssf?/cgi-free/getstory_ssf.cgi?n0505_BC_NY—contamination&&news&newsflash-newyork-syr. Last accessed July 7, 2001.

Brown, DeNeen L. "Culture Corrosion in Canada's North; Forced Into the Modern World, Indigenous Inuit Struggle to Cope." *Washington Post*, July 16, 2001, A-1.

Darvill, T., Lonky, E., Reihman, J., Stewart, P., Pagano, J. (2000). Prenatal exposure to PCBs and infant performance on the Fagan Test of Infant Intelligence, Neurotoxicology, 21, 1029-1038.

Dobbyn, Paula. "Contaminated Game has Natives Worried." Anchorage *Daily News*, May 2, 2000. http://www.ienearth.org/food_toxic.html. Last accessed July 6, 2003.

References

Fitzgerald, Syni-An Hwang, Brian Bush, Katsi Cook, and Priscilla Worswick. 1998. "Fish Consumption and Breast Milk PCB Concentrations among Mohawk Women." History of the Akwesasne Task Force on the Environment and Environmental Injustice. Wikipedia. Last accessed March 25, 2022. https://en.wikipedia.org/wiki/Akwesasne_Task_Force_on_the_Environment

Hoover, Elizabeth. *The River is In Us: Fighting Toxics in a Mohawk Community.* Minneapolis: University of Minnesota Press, 2017.

Hansen, Terri. "Akwesasne Mohawk Youth Are Still at Risk of Industrial Pollutants." Indian Country Today Media Network, June 20, 2011. http://indiancountrytodaymedianetwork.com/2011/06/20/akwesasne-mohawk-youth-are-still-risk-industrial-pollutants-39158. Last accessed October 4, 2015.

Hansen, Terri. "Akwesasne Mohawk Youth Are Still, at Risk of Industrial Pollutants." *Indian Country Today,* September 13, 2018. https://indiancountrytoday.com/archive/akwesasne-mohawk-youth-are-still-at-risk-of-industrial-pollutants. Last accessed May 15, 2022.

Hulen, David. "Hunt is on for Pollutant Traces in Bering Sea; Alaska Villagers, Scientists Wonder if Toxic Substances are Endangering Animals and people Who Eat Them." *Los Angeles Times*, August 15, 1994, A-5.

Jacobson, J. L., Jacobsen, S. W., Humphrey, H. E. B. (1990). Effect of in utero exposure to PCBs and related compounds on cognitive functioning in young children. Journal of Pediatrics, 116, 38-45. https://digitalcommons.usu.edu/kicjir/vol5/iss1/1. https://doi.org/10.26077/4f19-r6064. Last accessed February 5, 1993.

Jacobson, J. L. and Jacobson, S. W. (1996). Intellectual impairment in children exposed to polychlorinated biphenyls in utero. *The New England Journal of Medicine*, 335, 783–789.

Jacobson, S. W., Fein, G. G., Jacobson, J. L., Schwartz, P. M., Dowler, J. K. (1985). The effect of intrauterine PCB exposure on visual recognition memory. Child Development, 56, 853-860.

Johansen, Bruce E. *Life and Death on Mohawk Country.* Golden, Colo.: North American Press/Fulcrum, 1993.

Kader, Charles. "Outrage Builds in Akwesasne over PCB Contamination Reality." *Two Row Times*, February 12, 2014, p. 1. https://tworowtimes.com/news/regional/outrage-builds-akwesasne-pcb-contamination-reality/. Last accessed March 30, 20126.

Krook, L. and G.A. Maylin. "Industrial Fluoride Pollution: Chronic Fluoride Poisoning in Cornwall Island Cattle." Cornell *Veterinarian* 69(Supplement 8):1-70 (1979). http://www.ncbi.nlm.nih.gov/htbin-post/Entrez/query?uid=467082&form=6&db=m&Dopt=r. Last accessed December 4, 1985.

McRae, Kim Ellen, "Effects of PCB Contamination on the Environment and the Cultural Integrity of the St. Regis Mohawk Tribe in the Mohawk Nation of Akwesasne" (2015). *Graduate College Dissertations and Theses.* 522. https://scholarworks.uvm.edu/graddis/522. Last accessed January 27, 2016.

"Mohawk Government Opposes Grasse River Cleanup Plan." *Watertown Daily Times*, October 2, 2012. http://www.watertowndailytimes.com/article/20121002/NEWS09/710029710

"Mohawk Man Arrested for Taking Backhoe to Superfund Site." Indian Country Today Media Network, August 12, 2011. http://indiancountrytodaymedianetwork.com/article/mohawk-man-arrested-for-taking-backhoe-to-superfund-site-46891. Last accessed August 22, 2016.

"Mohawks Express Gratitude for Akwesasne Health Advocate" Indian Country Today Media Network, January 6. 2012. http://indiancountrytodaymedianetwork.com/article/mohawk-man-fights-to-remove-toxic-hazardous-waste-from-river-104947. Last accessed March 4, 2013.

"Mohawks Say EPA Alcoa-Superfund Cleanup Plan Falls Short." Indian Country Today Media Network, January 21, 2013. http://indiancountrytodaymedianetwork.com/2013/01/21/mohawks-say-epa-alcoa-superfund-cleanup-plan-falls-short-147131. Last accessed April 22, 2014.

Negoita, S., L. Swamp, B. Kelley, and D.O. Carpenter. "Chronic Diseases Surveillance of St. Regis Mohawk Health Service Patients." *Journal of Health Management Practice* 7:1(2001):84-91. https://www.ncbi.nlm.nih.gov/pubmed/11141627. Last accessed April 15, 2003.

Ravenscroft J, Schell LM; Akwesasne Task Force on the Environment. Patterns of PCB exposure among Akwesasne adolescents: The role of dietary and inhalation pathways. Environ Int. 2018 Dec;121(Pt 1):963-972. https://doi.org/10.1016/j.envint.2018.05.005. Epub 2018 Oct 22. PMID: 30359956. Last accessed May 11, 2019.

Seely, Hart. "Toxins Remain 18 Years Later: Landfill Near Massena Polluting Water where Mohawk Children Played." Syracuse *Post-Standard*, June 24, 2001, n.p.

Sengupta, Somini. "A Sick Tribe and a Dump as a Neighbor." *New York Times*, April 7, 2001. http://www.nytimes.com/2001/04/07/nyregion/07MOHA.html. Last accessed August 17, 2002.

Stewart, P. W., Reihman, J., Lonky, E. I., Darvill, T. J., Pagano, J. (2003). Cognitive development in preschool children prenatally exposed to PCBs and MeHg. *Neurotoxicology Teratology*, 25, 11-22.

Thomas, Katie. "Toxic Threats to Tribal Lands." *Newsday*, March 25, 2001. http://www.newsday.com/coverage/current/news/sunday/nd8399.htm. Last accessed July 5, 2001.

Chapter 3
The Deadly Yellow Powder

Until the late 1940s, radiation was unknown among the Navajo, and Pueblo in New. Mexico, and the Dene of Northwestern Canada. Cancers of the type caused by radiation were all but unknown. There was no word for it in Native languages. After World War II, uranium mining became a large industry in all of these places, and others, as stockpiles of nuclear weapons were accumulated as part of the "cold war" between the United States and the Union of Soviet Socialist Republics (USSR).

Native American men, many of them veterans of the war, arrived home without ready sources of employment, except in the new uranium mines. Many of them accepted jobs mining "yellow dust." What came next was a deadly, frequent, and very painful plague of cancers, birth defects, and other ailments that within four decades had killed several hundred Navajos, Pueblos, and Dene with radiation that can build quietly in the body, over decades and generations. It can cause multiple types of cancers, as well as birth defects, and other serious ailments.

3.1 The Human Price of Uranium among Native American Peoples

During 1950, Navajo sheepherder Paddy Martinez, a, brought a yellow rock into Grants, New Mexico, from nearby Haystack Butte. Word quickly spread in the area that the yellow rock was uranium, initiating a frenzy of mining from large underground networks to small "dog holes" that could accommodate only one man at a time. Boom. Navajo uranium miners at first hauled the radioactive uranium ore out of the mines like coal. Some ate lunch among the yellow rocks and drank water that flowed within the mines.

Families built homes (hogans) of radioactive earth as their sheep grazed on plants that grew out of small ponds that formed near the entrances to mines. Others grazed on radioactive tinged soil. On windy days, dust from uranium waste-tailings piles

© The Author(s), under exclusive license to Springer Nature Switzerland AG 2023
B. E. Johansen, *Resource Devastation on Native American Lands*,
https://doi.org/10.1007/978-3-031-21896-5_3

blew through villages, covering everything with yellow dust. For many years, no one told the Navajos, or other Native peoples, that the rocks or the dust would kill many of them within 10 to 20 years. It was a dirty, dusty job that paid well, compared to other alternatives. However, after several years, nearly all Navajo hogans in the area harbored the dust of death.

Almost half of the United States' recoverable uranium lies in New Mexico—and about half of that is situated within mining depths under the Navajo Nation. No one told Navajo families that within a few years, the vast majority of their homes would register on Geiger counters, a sign of pervasive and possibly lethal radiation poisoning. No one thought telling the people who were now living on poisoned earth was worth their time. The mine owners were not environmental crusaders. They were in this land that they considered bleak to make money. The Navajo were useful like donkeys might be, as beasts of burden.

Very few mining companies provided ventilation in the beginning of the boom. Some miners worked up to 20 hours a day in the mines, starting just after sandstone blasting of sandstone filled the mines with silica dust. Many mine owners did not even provide toilet paper to the miners. Miners who needed to relieve themselves did so with fistsfull of radioactive "yellowcake"—uranium ore.

The Kerr-McGee Company, the first to mine uranium on the Navajo Nation (in 1948), found the location very profitable, without taxes (at that time), nor pollution, health, or safety restrictions. They had no union to bargain for decent wages and benefits, so the veterans' labor was cheap, compared to that of imported workers. At the same time, uranium was in high demand for growing inventories of nuclear armaments, was expensive, the perfect corporate atmosphere for high profits.

On the reservation, nearly all the workers were Navajos, who were sent into shallow tunnels within minutes of blasting. They threw radioactive rocks into wheelbarrows and emerged from a growing network of shallow mines spitting black mucus. They coughed with a force that gave them headaches. Miners who worked under these conditions were exposed to 10–1,000 times the radon that later would be determined as safe. At the time, there were no standards for radon toxicity and, even if such existed, no one was measuring it.

In the National Institutes of Health's *Environmental Health Perspectives*, Carrie Arnold wrote (2014) that the miners and their families "were not told that the men who worked in the mines were breathing carcinogenic radon gas and showering in radioactive water, nor that the women washing their husbands' work clothes could spread radionuclides to the rest of the family's laundry after they had worked in the 521 now-abandoned uranium mines on the reservation ranging in size from 'dog holes' that could accommodate only a single man to large mines from which radioactive ore was extracted in carts on rails" (Arnold 2014).

Arnold also wrote that health workers were not allowed to interview uranium miners only after they agreed not to tell miners of potential health hazards that they encountered every day. Realizing that sharing information was the only way to convince government regulators to improve safety in the mines, the researchers accepted this restriction. The PHS monitored the health of more than 4000 miners between 1954 and 1960 without telling them of the threat to their lives. By 1965, the investigators reported that cumulative exposure to uranium had a positive correlation with lung cancer among white miners and had definitively identified the cause as increased radiation exposure. The effects of radiation exposure had been no secret among scientists even as early as 1950, when government workers monitored radiation levels in the mines that were as much as 750 times the limits deemed acceptable at that time. No one told the miners, however, and the miners did not read scientific journals or government reports.

3.2 The Scope of Uranium Mining on Native Lands

The digging of uranium mines on Native lands affected not only the Navajo and Laguna lands, according to Mary F. Calvert of the Pulitzer Center (2021): "After the invention of atomic weapons in 1945 and the subsequent development of nuclear power plants, mining companies dug more than 4000 uranium mines across the Western U.S. Although other tribes were affected—including the Hopi, the Arapaho, the Southern Cheyenne, the Spokane and [Secretary of the Interior under President Joe Biden) Deb] Haaland's own Laguna Pueblo—roughly 1000 of these claims were located on the Navajo Nation, which encompasses 27,000 square miles where Arizona, Utah and New Mexico meet. Over the next four decades, miners contracted by the U.S. government blasted 30 million tons of ore out of Navajo land with little environmental, health or safety oversight. Eventually, demand declined, deposits played out and the pits were abandoned. Much of the damage, however, had already been done." (Calvert 2021). Laguna Pueblo's open-pit Jackpile-Paguate Mine was among the world's largest.

When the mines were active, companies recruited Navajo men to work them, and hired women and children as support staff, wrote Calvert. To save money and avoid leaving a paper trail, they "paid" them with sacks of sugar, flour, potatoes, and coffee. Exposure to radioactive ore and toxic by-products such as arsenic, cadmium and lead was commonplace—a fact of everyday life. Navajo families crushed the poisonous rock to make concrete. Homes were built from abandoned mine tailings. Kids played on waste piles. Herders watered sheep in open, unreclaimed uranium pits. Husbands came home covered in uranium dust; wives washed their clothes. Everyone drank the contaminated groundwater; everyone inhaled mine dust borne on the hot, dry desert winds. They still do today.

"They never told us uranium was dangerous," said Cecilia Joe, 85, a Navajo woman who worked as a miner from May 1949 to June 1950. The federal government had studied and documented the danger in depth, but deliberately kept it secret. "We washed our faces in it. We drank in it. We ate in it. It was sweet...." Her neighbors, Joe continued, stopped working at the mine and moved to Utah when their daughter Mary was 2. "The little girl passed away while they were there. She started coughing up blood. She began having convulsions. They say that she twisted into unnatural positions....When Joe was 3, seven of her siblings died in a span of 20 days.... Three of her brothers started coughing up blood" (Calvert 2021).

3.3 Uranium from Russia: An Alarm for U.S. Native Peoples?

Just as Native nations inside the United States have been turning away from mining and milling of a lethal uranium-related legacy because many Native people have died, the U.S. federal government has become concerned that events in other countries (the case in question being the Russian invasion of Ukraine) could cause the United States to call for a rapid increase in production of "domestic" uranium to fuel more than 90 nuclear power plants. In 2022, the United States purchased almost half of its uranium from Russia and two of its former territories in the Soviet Union, Kazakhstan, and Uzbekistan.

During the Russian invasion of Ukraine, the United States imposed sanctions on Russia involving every major fuel source except uranium, staving off a confrontation over such questions as: under what conditions would federal law force Native nations to open their uranium mines? Given the number of Native American lives that have been lost to uranium milling and mining, (discussed below) plus the implications of forced mining for Native sovereignty, resulting confrontations could be tumultuous.

"With some of the most coveted uranium lodes found [on or]. around Indigenous lands, the moves are setting up clashes between mining companies and energy security hawks on one side and tribal nations and environmentalists on the other," wrote Simon Romero for the *New York Times*. That confrontation could pose an important test for almost two centuries of American Indian law that reaches back to the bench of Chief Justice John Marshall's U.S. Supreme Court which, during the 1820s and 1830s, defined Native lands as semi-sovereign domestic, dependent nations.

As will be discussed below, the Navajo (Dine) have made uranium mining illegal on their lands, the largest Native reservation in the United States. In other areas, such as the homeland of the Havasupais near the Grand Canyon "Whose people," wrote Romero, "have lived in the canyonlands and plateaus of the Grand Canyon since time immemorial," call the area of the mining site *Mat Taav Tiijundva*—"Sacred Meeting Place" in a rough translation. Following Russia's invasion, wrote Romero. "Mining executives have seen the site as something else: the tip of the spear in their jostling to advance American uranium projects."

"We have the ability to reduce dependence on Russian uranium right now," said Mark Chalmers, chief executive of Energy Fuels, a Colorado company whose interests in the area have provoked a legal offensive to increase uranium production (Romero 2022). The Grand Canyon area experienced something of a uranium mining boom between about 1955 and 1985. A mine built during that time spilled an enormous amount of arsenic and uranium-tainted water on land that contains Havasupai grave sites and imperiled their water quality. The mine was never fully operational after that, and generally advocates of uranium's mining and milling have not been popular around the reservation.

However, increasing U.S. uranium production has never been more popular among its lobbyists in Washington, D.C. "A robust domestic supply chain for nuclear fuel has never been more important for our nuclear fleet," Scott Melbye, the president of the Uranium Producers of America, said in testimony before the U.S. Senate during March, 2022. "The sooner we decouple our nuclear industry from Russia, the sooner Western nuclear markets can get to work to fill the gap" (Romero 2022).

The United States, as of 2022, produced about one-fifth of its electricity with uranium. Many Havasupais and their allies are steadfastly against contributing more radioactivity to their lives. "The possibility that this mine could go forward is unthinkable under an administration [under President Joe Biden] that has made promises to prioritize environmental justice," said Amber Reimondo, energy director at the nonprofit Grand Canyon Trust. One look at the death toll and more than 400 uranium-related deaths as well as hundreds of small, abandoned uranium mines afflicting the Navajo, the Laguna Pueblo, and other nearby Native nations is enough to convince most of the Havasupai that any new mining is a physical and spiritual dead end.

Many of the Havasupai believe the mining site to be "sacred, dotted with burial places and remains of homes and sweat lodges," said Carletta Tilousi, a former Havasupai Tribal Council member who has been fighting the mine for decades (Romero 2022). The fact that mining pierced an aquifer is seen as a warning among many attentive Havasupai also have been watching the death and devastation wreaked by the mining of uranium among the Navajo. In the meantime, the uranium lobby has been slathering over uranium prices, up about 30% in less than two months after Russia invaded Ukraine (as of mid-May, 2022).

David Kreamer, a professor of hydrology and an authority on groundwater contaminants at the University of Nevada, Las Vegas, said that the Havasupai site is a potential "time bomb," noting that the aquifer piercing "released millions of gallons of water high in both uranium and arsenic" (Romero 2022).

Noting the speed with which permits are been issued for the new mine, Ms. Tilousi said: "The domestic uranium industry is set to go into overdrive again, but we [Havasupai opponents] will lie down in front of the mine's entrance to keep it from fully functioning if we have to. We'll make them understand this is about much more than money" (Romero 2022).

Some Native peoples were close to "ground zero" for the earliest nuclear tests nearby. In addition, the Navajo Nation was the major site for the largest nuclear accident in the history of the United States, worse than the much more intensely publicized event at Three Mile Island, during the same year. The Three Mile Island event was cleaned up within 16 years, which is warp speed in the Nuke repair business. In 2022, 45 years post-spill, the Navajo accident, which involved a flood containing several million tons of radioactive water and mud, had not been cleaned except in the most minimal of ways.

On Navajo land, uranium's mining was deemed illegal after 2005, as the Navajos people heeded their "original instructions"—a code of behavior requiring that evil substances remain in the ground. After roughly 45 years, the U.S. Environmental

Protection Agency (EPA) was still planning how to "remediate" about 500 uranium mines on or near the Navajo Nation." In 45 years, bureaucracy and corporate willingness to suck up every available dollar in the sacrosanct realm of nuke repair had contracted a perpetual-motion money-devouring machine that became very good at telling the public and government overseers that they were on the road to remediation, and then doing very little except (as the companies proudly reported whenever contract renewal came around) creating a little work for some small-scale Navajo businesses.

3.4 Navajo Uranium Fuels U.S. Arsenal

The Navajos' negative experience with uranium goes back for decades before the conflicts of 2022. Uranium mined from Navajo and other Native American lands supplied a substantial proportion of the fuel for early nuclear power plants as well as the U.S. nuclear arsenal. By the 1970s, many of the earliest miners were dying of lung cancer. In Washington State, nuclear waste from the Hanford plants afflicted the Yakamas. Uranium mining also has continued to cause a plague of several types of cancer at the Laguna Pueblo in New Mexico. The Navajos succeeded in stopping uranium mining and milling only after several hundred people had died of its cancerous effects and many more had suffered the tortures of cancers that once were nearly unknown in their country.

Larry King, an activist and former uranium mine worker, who is Dine said that he had mined uranium in his street clothes. "So it was just usually one of my old shirts, my pants. No gloves. No respirator. Nothing. So everybody's breathing all that dust." Another former uranium worker, Linda Evers, said she wasn't told about the dangers associated with uranium exposure. "When we had safety meetings, it was about regular first aid," she said. "There was no mention of radiation—or any side effects from it" (Gounder 2022).

By 2013, about 400 former Navajo uranium miners had died from maladies mainly caused by exposure to radiation, according to Chris Shuey, an environmental health researcher with the Southwest Research and Information Center (Knight 2013). Not all of the miners who died worked in the mines themselves. Many others died from a wide variety of malignant cancers because they lived with the "yellow dirt" in windswept waste (tailings) piles that blew into every crack, outdoors and indoors, for several decades.

In addition to mining sites, New Mexico provided the United States with the first site for an atomic bomb test. It was detonated on a 100-foot tower on July 16, 1945. The Trinity site, located on what today's U.S. Army White Sands Missile Range, is surrounded by several Native American communities: 19 American Indian pueblos, two Apache reservations, and several Navajo towns. Observers, both Native peoples and U.S. government witnesses said that ash fell lightly for several days after the test. However, the effects of this fallout were not investigated until 2014, when the National Cancer Institute started study of radiation levels in New Mexico from that first test blast. With no evidence of effects, none of the people who observed the blast were eligible for compensation under the Radiation Exposure Compensation Act, which covers other nuclear and uranium workers as well as "down-winders" who were exposed to radiation from later atomic tests (Lee 2014).

3.5 Radioactive from the Inside Out

Almost 100% of the rock that the mines produced was waste, "tailings" dumped near mine sites after the uranium had been extracted. One mesa-sized waste pile grew to about a mile long and roughly 70 feet high. When wind was strong, tailings dust blew into peoples' homes, settling into water supplies. During the 1950s, the Atomic Energy Commission (AEC) told anxious residents that the dust harmless. The lack of warnings was a public-relations stunt. The AEC had known that uranium was lethal for years.

Even so, no one considered environmentally appropriate ways to deal with the tailings piles. Even if the tailings were to be buried—a staggering task—radioactive leaks would leak into the surrounding water table. A 1976 Environmental Protection Agency (EPA) report found radioactive contamination of drinking water on Navajo land under the Grants, New Mexico, area near a uranium mining and milling facility (Eichstaedt 1994, 208).

Arnold (2014) wrote of the mine tailings' legacy more than half a century after the first uranium was mined: "In a low, windswept rise at the southeastern edge of the Navajo Nation, Jackie Bell-Jefferson prepares to move her family from their home for a temporary stay that could last up to seven years. A mound of uranium-laden waste the size of several football fields, covered with a thin veneer of gravel, dominates the view from her front door. After many years of living next to the contamination and a litany of health problems she believes it caused, Bell-Jefferson and several other local families will have to vacate their homes for a third round of cleanup efforts by the U.S. Environmental Protection Agency."

Members of Bell-Jefferson's family had used radioactive water for years, "splashing and swimming in pools of radioactive water that had been pumped out of the mines [onto] their property. The contaminated water looked and tasted perfectly clean. Families used it for cooking, drinking, and cleaning. Hogans and corrals were built with mine wastes, as were roads" (Arnold 2014).

Gilbert Badoni of Shiprock, New Mexico, on the Navajo reservation, said that as a child, he played in the tailings with siblings and drank radioactive water during the decades his father worked in uranium mines in Colorado during the 1950s and 1960s. On some days, dry winds blew gritty plumes of dust from tailings piles through the streets of many Navajo communities. "We used to play in it," said Terry Yazzie of an enormous tailings pile behind his house. "We would dig holes and bury ourselves in it" (Eichstaedt 1994, 11). The neighbors of this tailings pile were not told that it was dangerous until 1990, 22 years after the mill that had produced the enormous pile had closed and 12 years after the. U.S. Congress had authorized the cleanup of uranium mill tailings in Navajo country. Abandoned mines also were used as shelter by animals that inhaled radon and drank contaminated water. Local people drank milk from contaminated cows and ate their contaminated meat.

Despite the lack of official attention, residents were adding things up. One said that his father had arrived at home after work covered in yellow dust. After they told officials about this again, the residents understood that the dust was radioactive and therefore a cancer-causing poison. The dust which covered everything in their small home as their mother brushed it off his clothes. He already had lung cancer, and blamed his lung problems and his siblings' cancers for that exposure. "The U.S. government has abused innocent women and children. They have abused my family," he said, choking back tears. "They have abused my Navajo people. That's not right" ("Uranium Miners" 2000).

Uranium mine dust produced silicosis in the miners' lungs in addition to lung cancer and other problems associated with exposure to radioactivity. By the 1960s, at least 200 miners had died of uranium-related causes. That number had doubled by 1990. Radioactivity also contaminated drinking water in parts of the reservation, producing birth defects and Down syndrome, both of which had been previously unknown among Navajos. The victims of radiation poison were filling local hospitals. Others died at home. Of all infant deaths in areas served by the Navajo Indian Health Service area for the years 1990 through 1992, 35% were caused by congenital anomalies. Mortality attributed to malignant neoplasms (at an age-adjusted rate) was 78.5% in the years from 1990 through 1992 (Colomeda 1998). A 1976 EPA report had found radioactive contamination of drinking water in the Grants area on the Navajo reservation near a uranium mining and milling facility.

The U.S. government had known of the radioactive risk at least since 1978, as the Department of Energy had released a Nuclear Waste Management Task Force report which indicated that people living near tailings piles had at least twice the risk of lung cancer in the general population. Even then, the Coalition for Navajo Liberation was aware that miners were dying of lung cancer and other uranium-related illnesses. As Navajo miners continued to die, children who played in water that had flowed over or through abandoned mines and tailings piles came home with burning sores. Downwind of uranium processing mills, dust from "yellowcake" (enriched uranium that may be used in the manufacture of nuclear fuel) sometimes was so thick that it stained the landscape half a mile away.

Peter Eichstaedt wrote that miners watched members of their families die because radiation poisoning permeated their entire lives. Some miners were put to work packing hundred-pound barrels of yellowcake. Some of the miners ingested so much of the dust that it was "making the workers radioactive from the inside out" (Eichstaedt 1994, 11). And what were the Navajos in the area to do? Quit drinking water from their wells? Quit eating their meat animals? Hold their breaths on windy days? Sell their hogans on radioactive land at vastly deflated prices and move away?

3.6 Opposition to Uranium Mining

Opposition the long-term poisoning of land, water, and air, and water by low-level radiation has become a major concern among non-Indians and Indians. Beginning during the 1970s, these conditions provoked demands for a moratorium on all exploration, mining, and milling of uranium until issues related to radioactive tailings and other problems related to waste disposal problems have been solved. Doris Bunting of Citizens Against Nuclear Threats, a mainly non-Native group that joined with the Coalition for Navajo Liberation (CNL) and the National Indian Youth Council (NIYC), to oppose uranium mining, supplied information indicating that contaminated sediments had spread into the Colorado River basin, a major source of water for a large proportion of people and industries in the area on and near the Navajo Nation and nearby areas.

The prospect of death or serious illness haunted hundreds of former Navajo uranium miners. By the end of 1978, more than 700,000 acres of Native American land were under lease for uranium exploration with prospects for development, notably mining, in an area centering on Crownpoint and Shiprock, both on the Navajo Nation. Atlantic Richfield, Homestake, Continental Oil, Humble Oil, Kerr-McGee, Mobil Oil, Pioneer Nuclear, Exxon, and United Nuclear were planning to mine, or already extracting ore. During the 1980s, the mining frenzy subsided somewhat as the area was affected by recession, and slowing demand because the nuclear arms race had showed, reducing demand.

Even as the uranium boom ended, many of the miners who had been condemned to slow death by lung cancer did not yet know that the yellow ore they had mined was killing them. Peter Eichstaedt's painstakingly detailed study described how the miners learned what had been done to them and how they went about winning at least the possibility of compensation from the U.S. government. Using a combination of interviews and scientific data, Eichstaedt proved that the U.S. government (particularly the AEC) knew that uranium mining was poisoning the Navajos from the beginning, an important point in the debate about compensation that by that time had resulted in a small proportion of the miners collecting $100,000 each as compensation. Government agencies mining companies were very tight-fisted with both information from medical records and testimonies from miners out of concern for national security and profits.

3.7 "Original Instructions," Uranium Mining, and Navajo Cultural Values

The mining of uranium became a major cultural and political issue among the Navajo during the 1970s as the death toll from several types of cancers rose, along with other maladies. By this time, Lung cancer had become so widespread and wrenchingly evident that many Navajo began to reexamine their traditional teachings about the need for harmony with regard to the Earth. Following debate centered on this examination, uranium mining and milling in 2005 became illegal in the Navajo Nation.

As uranium's human toll was released to the public, the CNL pressed to have uranium mining and milling outlawed on the reservation, a position that eventually prevailed in the Navajo government after a quarter of a century of debate and industry lobbying. Esther Keeswood, from Shiprock, a member of the CNL, a reservation town near several tailings piles, in 1978 said that the CNL had documented the deaths of at least 50 Shiprock residents (including several uranium miners) from lung cancer and related diseases. The number of disclosed deaths increased after that so that by about 1995, it was reaching about 400.

Uranium's yellow dirt comprises a part of the natural order of things in Navajo Country. A ban on its mining and milling was regarded by traditional people as a cultural victory, a return, at least in part, to the original instructions of creation. The yellow dirt in mythology is believed to be the antithesis of the life-giving yellow powder of corn pollen:

> In one of the stories the Navajos tell about their origin, the *Dineh* (the people) emerged from the third world into the fourth and present world and were given a choice. They were told to choose between two yellow powders. One was yellow dust from the rocks, and the other was corn pollen. The Dineh chose corn pollen, and the gods nodded in assent. They also issued a warning. Having chosen the corn pollen, the Navajo were to leave the yellow dust in the ground. If it was ever removed, it would bring evil. (Eichstaedt 1994, 47)

In this cultural interpretation, the commercial use of the yellow dirt turned it into *Leetso*, a powerful monster that inflicted punishment on the Navajo people in the form of disease and suffering. The only way to restore harmony with nature (a very important attribute in Navajo culture) is *hozho nashaadoo* ("to walk in harmony") (Csordas 1999). The only proper response was to restore harmony by banning the mining and milling of uranium. The yellow dirt is a poison that disrupts the gathering of sacred herbs in contaminated areas. The Navajo seek to balance elements of land, water, and sunlight (fire) (Woody et al. 1981). Some of the elders blamed themselves for allowing the disruption of *hozho nashaadoo*.

3.8 The Largest Uranium Spill in the United States

The United States' largest nuclear accident (that is, expulsion of radioactive material) began at 5 A.M. on July 16, 1979, on the Navajo Nation, not at the Three Mile Island plant in Pennsylvania. On that morning, several million tons of uranium mining wastes—tailings mixed with water—gushed through a breached

3.8 The Largest Uranium Spill in the United States

packed-mud dam near Church Rock, New Mexico. With the tailings, 94 million gallons of radioactive water flowed through the dam before the crack was repaired.

"By the end of the week, more than 140,000 nearby residents left their homes," said Céline Gounder in her podcast (2022). No one was killed, at least not initially. The accident was all over the news. Even President Jimmy Carter visited the plant for a briefing on the accident a few days later.

Just a few months later, in July 1979, the Navajo accident released *three times* the amount of radioactive material as Three Mile Island. That time, the news barely covered it at all. This spill occurred at a mine in Church Rock, New Mexico—a small town on the Navajo Nation. A Diné uranium mine worker named Larry King, who watched the dam break, saw part of roughly 94 million gallons of toxic wastewater rush into surrounding fields. "The acidity of that wastewater was equal and above the acidity of battery acid," said King. Gounder added that these toxins did not stay in Church Rock, but flowed downstream into Gallup, N. M. (2022).

An official from UNC, the parent company of the mill where the spill had begun, later told a U.S. House of Representatives committee that because of rough terrain, cleanup was impeded by rough terrain, and had to be done by hand. During that same hearing, an environmental analyst noted that UNC initially assigned 10 men to the cleanup. The analyst said the workers were given only shovels and 55-gallon barrels to remove the contaminated sediment from the Puerco River. (Gounder 2022).

King added that "All I can remember was seeing the laborers, in wetsuits, shoveling all that slime off of the embankments of the wash and scooping them into containers" (Gounder 2022). Some of the radioactive sand that had been shovelled into barrels tipped over in rainstorms. Residents were furious. Had it happened in Phoenix or Los Angeles, "you better believe there'd be a lot of action done," many said (Gounder 2022).

This is worth repeating, *because of the racism it implies: the Three Mile Island's cleanup was finished in 1993. In 2022, The Church Rock Navajos were still looking at about 500 abandoned mines, a few of which had been partially cleaned on a trial basis—the U.S. government had not passed the testing stage in most of the contaminated mines in 45 years, as of 2022!*

The Church Rock mill site had not been even partially remediated by that tomer. King thinks he knows why. "Because we're an Indigenous community. We're a poor community" (Gounder 2022).

Along the Rio Puerco in Navajo Country, the contaminated floodwater moved quickly. By 8:00 A.M., radioactivity was monitored in Gallup. The contaminated Rio Puerco registered 7000 times the allowable standard of radioactivity for drinking water below the broken dam shortly after the breach was plugged, according to the Nuclear Regulatory Commission (NRC). A few newspaper stories said that it was sparsely populated and that the spill "posed no immediate health hazard" (Gounder 2022). Even The New Mexico Environmental Improvement Division asserted that the spill had been only "potentially hazardous … its short-term and long-term impacts on people and the environment were quite limited" (Johansen 1997, 11). With these soothing words, the same reports also recommended that ranchers in the area avoid watering their livestock in the Rio Puerco. After the Rio Puerco spill, however, several Navajos said that calves and lambs had been born without limbs as well as other severe birth defects. Other livestock developed sores, became ill, and died after drinking from the river.

The same report noted that the river water was not being used for human consumption and that "the extent to which radioactive and chemical constituents of these waters are incorporated in livestock tissue and passed on to humans is unknown and requires critical evaluation" (Johansen 1997, 12). The report also said that the accident's effect on groundwater should be studied more intensely. Tom Charley, a Navajo, told a public meeting at the Lupton Chapter House that "The old ladies are always to be seen running up and down both sides of the [Rio Puerco] wash, trying to keep the sheep out of it" (Johansen 1997, 12). The Centers for Disease Control (CDC) examined a dozen dead animals and called for a more complete study in 1983, then dropped the subject.

Even as the Rio Puerco ran radioactive, the town of Grants styled itself as the "Uranium Capital of the World" as new pickup trucks appeared on the streets and mobile home parks grew around town, filling with non-Indian workers who were taking part in the uranium boom. For several years, before the boom abruptly ended in the early 1980s, many workers in the uranium industry earned $60,000 or more a year (about $200,000 in 2022 dollars). Following a collapse of demand for uranium in the early 1980s, however, Grants dropped the nickname "Uranium Capital of the World" and began promoting itself as a haven for retirees under the new slogan "Grants Enchants."

3.9 The Struggle for Compensation

By 1978, the Navajos were beginning to trace the roots of a lung cancer epidemic that had perplexed many of them, because the disease had been very rare among them before World War II. Harris Charley, who worked in the mines for 15 years, told a U.S. Senate hearing in 1979, "We were treated like dogs. There was no ventilation in the mines." Pearl Nakai, daughter of a deceased miner, told the same hearing, "No one ever told us about the dangers of uranium" (Grinde Jr. and Johansen 1995, 214). The Senate hearings were convened by Senator Pete Domenici, Republican of New Mexico, who was seeking compensation for disabled uranium miners and for the families of the deceased. "The miners who extracted uranium from the Colorado Plateau are paying the price today for the inadequate health and safety standards that were then in force," Domenici told the hearing, held at a Holiday Inn near Grants (Grinde Jr. and Johansen 1995, 214).

U.S. Senate hearings during 1979 initiated proposals to compensate the miners for what investigators called deliberate negligence. By 1930, radioactivity in uranium mines had been associated with lung cancer by tests in Europe Scientific evidence linking radon gas to radioactive illness existed after 1949, but measures to ventilate the Navajo mines were never implemented, as the government pressured Kerr-McGee and other producers to increase the amount of uranium they were mining. The PHS recommended ventilation in 1952, but the AEC said that it bore no responsibility for the mines despite the fact that it bought more than 3 million pounds of uranium from them in 1954 alone.

Dr. Joseph Wagoner, special assistant for occupational carcinogens at the Occupational Safety and Health Administration, a federal agency, said that of 3500 persons who mined uranium in New Mexico, about 200 had died of cancer by the late 1970s. In an average population of 3500 persons, 40 such deaths could be expected. The 160 extra deaths were not the measure of ignorance, he said. Published data regarding the dangers of radon were widely available to scientists in the 1950s, according to Wagoner. Health and safety precautions in the mines were not cost-effective for the companies, he said. "Thirty years from now we'll have the hidden legacy of the whole thing," Wagoner told Molly Ivins (1978, n.p.).

For a dozen years, bills that would have compensated the Navajo miners were introduced, discussed, and left to languish and die in congressional committees. After years of false starts, compensation was finally approved after more than a decade of congressional debate. Next, the federal bureaucracy stood in the way. By 1990, the cancer-related death toll among former miners had risen to 450 and was still rising. By the early 1990s, about 1100 Navajo miners or members of their families had applied for compensation related to uranium exposure. The bureaucracy had approved 328 cases, denied 121, and withheld action on 663, an approval rate that Representative George Miller, chairman of the House Natural Resources Committee, characterized as "significantly lower than in other cases of radiation compensation" (Johansen 1997, 12). Representative Miller said that awards of compensation were being delayed by "a burdensome application system developed by the Department of Justice" (Johansen 1997, 12).

3.10 Delays in Compensation

By the beginning of 2001, former Navajo uranium miners and millers, as well as their families, were holding meetings and protest marches in a renewed attempt to collect compensation under the Radiation Exposure Compensation Act for cancers associated with radiation exposure. Meetings and protest marches were held in Grants; in Cortez, Colorado; and in or near several towns on the Navajo Nation. National legislation compensating victims of uranium mining and fallout had been enacted in 1990, guaranteeing as much as $100,000 to miners and down-winders who had suffered radiation exposure during nuclear weapons tests. The program was not funded until 1992 and even then only at very low levels. It ran out of money during 1993, when the government began issuing IOUs.

During July 2000, President Bill Clinton signed an amendment to the 1990 law extending its provisions to millers and transporters of radioactive uranium ore, once again without adequate funding to implement the law's provisions. In the meantime, increasing numbers of former uranium miners were suffering illnesses and deaths due to their exposure. The amendments expanded the list of eligible afflictions and reduced the amount of time a miner was required to have spent working with uranium to be eligible for the compensation program. The measure also opened the program to those who had worked in open-pit uranium mines and uranium milling plants as well as underground mines. Many miners had complained that they were excluded by the overly stringent rules of the original law.

Melton Martinez, who has been instrumental in the struggle to gain compensation for miners and millers, said the government knew that uranium had caused many cancers and organ failures. Even wives of workers who washed the yellow-coated clothes of their miner husbands were experiencing cancers, Martinez said (Purdom 2001). Mayor Elisabeth Lopez-Rael of Milan, New Mexico, told a gathering there that she knew a man in Grants who was given an IOU by the government. "He told me he probably wouldn't live long enough to spend it," Lopez-Rael said (Purdom 2001).

Several former miners raised the possibility that aquifers in the area remained contaminated by radioactivity from the former uranium mines. One man asked, "Where does the contamination go?" "Into the water," they were told. Because of radioactivity's persistent nature, Martinez said, "we are doing this not only for ourselves and our neighbors; we are doing this for our children too" (Purdom 2001). "They did experiments on us like guinea pigs. It makes me angry," one former miner said as he sat on the steps outside the U.S. House of Representatives at the U.S. Capitol. "I would have lived longer, but they gave me a shorter life on this Earth" ("Uranium Miners" 2000).

3.11 Cleanup Planned at a Few Spill Sites after Three Decades

After almost three decades of appeals from the Navajo Nation, the EPA approved plans in September 2011 to clean up the Northeast Church Rock Mine near Gallup, the largest abandoned uranium mine in Navajo Country. The mine was used between 1967 and 1982 by the United Nuclear Corporation, which left behind roughly 1.4 million tons of uranium and radium-contaminated soil near its 125-acre uranium previously described site of the largest release of radioactive waste in the United States on July 16, 1979.

Cleanup of the site was expected to require several years (no one knows exactly *how* long). According to an Environment News Service report, cleanup workers "will place the contaminated soil in a lined, capped facility employing the most stringent standards in the country. When complete, the cleanup will allow unrestricted surface use of the mine site for grazing and housing, according to the EPA." People in the area have been at risk "from inhaling radium-contaminated dust particles and radon gas or utilizing contaminated rainwater and runoff that has pooled in the ponds. There is an elevated risk associated with livestock that may graze and [drink] water on the site." Radium exposure at high levels over several

3.11 Cleanup Planned at a Few Spill Sites after Three Decades

years, according to the EPA, "can result in anemia, cataracts, and cancer, especially bone cancer, and death" ("Cleanup" 2011).

"This is an important milestone in the effort to address the toxic legacy of historic uranium mining on the Navajo Nation," said Jared Blumenfeld, administrator for the Pacific Southwest Region. "This plan is the result of several years of collaboration between EPA, the Navajo Nation, and the Red Water Pond Road community living near the mine." "On behalf of the Navajo Nation, I appreciate the efforts of the US EPA and Navajo EPA" ("Cleanup" 2011).

In 2022, most of the 500-plus abandoned mines that had been known to have been flooded during the late 1979 flood remained to be cleaned up, as the EPA was still awarding contracts in the $220 million range to do the work. The cleanup had turned into a perpetual motion money machine, pumping tax dollars into the pockets into a handful of companies. Witness the following EPA press release dated February 11, 2021 (EPA 2021).

February 11, 2021

SAN FRANCISCO—Today, the U.S. Environmental Protection Agency (EPA) announced three contract awards for cleanup efforts at more than 50 abandoned uranium mine sites in and around the Navajo Nation. The Navajo Area Abandoned Mine Remedial Construction and Services Contracts (AMRCS), worth up to $220 million over the next 5 years, were awarded to Red Rock Remediation Joint Venture, Environmental Quality Management Inc. and Arrowhead Contracting Inc.

"EPA continues to work with the Navajo Nation EPA and local communities to address the legacy of abandoned uranium mines," said Deborah Jordan, Acting Regional Administrator for the EPA's Pacific Southwest office. "These contract awards mark a significant step in this ongoing work."

Most of the funding for the contracts came from the nearly $1 billion settlement reached in 2015 for the cleanup of more than 50 abandoned uranium mine sites for which Kerr McGee Corporation and its successor, Tronox, had responsibility. In addition to the funds from that Tronox settlement.... The Navajo Nation [has] secured funding agreements, through enforcement agreements and other legal settlements, for the assessment and cleanup of approximately 200 abandoned uranium mine sites on the Navajo Nation. That is a $220 million price tag, with 50 of more than 500 mines cleaned, plus about 200 others assessed and prepared for cleaning. The work was planned for 5 years, with an option for extension, as some similar jobs had been

On the sunny side of the perpetual motion money machine, some of the money was earmarked for Navajo employment opportunities. "EPA worked closely with [the] Navajo Nation to develop contracts that incentivize creating employment opportunities for Navajo residents and building local economic and institutional capacity. The contractors selected are classified as small businesses, two of which are owned by Native Americans. In addition, the companies have partnered with other Navajo-owned firms to ensure Navajo communities benefit economically from the ongoing work to clean up their land. To further direct benefits from this cleanup investment to the Navajo communities affected by this legacy pollution, each company will develop training programs for Navajo individuals and businesses to promote professional growth."

And then the EPA patted itself on the back:

Cleanup of the abandoned uranium mines is a closely coordinated effort between multiple federal agencies and the Navajo Nation. During the Cold War, 30 million tons of uranium ore were mined on or adjacent to the Navajo Nation, leaving more than 500 abandoned mines. Since 2008, [the] EPA has conducted preliminary investigations at all of the mines, completed 113 detailed assessments, cleaned up over 50 contaminated structures, provided safe drinking water to over 3000 families in partnership with the Indian Health Service, and completed cleanup, stabilization or fencing at 29 mines.

More than 500 abandoned mines in 1977. Five hundred abandoned mines in 2022. That is 45 years of ignoring the problem and then, once the biggest radiation spill in U.S. history was officially deemed as being worth cleaning up by its status as a Superfund site decades before the first abandoned mine was ready to rejoin the ranks of clean earth. "Legacy pollution" has such a pleasant resonance to it. Another way to phrase might have been "remaining potentially radioactive, fatal earth."

Some other abandoned uranium mines also were being slowly cleansed, but only a small fraction of the several hundred that were left idle. One was the Skyline Mine. Jason Musante, an EPA coordinator for a cleanup costing $6 million, told Judsy Fahys of the *Salt Lake Tribune*, "I've got to make this hazard go away as soon as possible.... It's already been too long." Radioactive debris was being packed into plastic bins (which Musante described as "giant Tupperware"), then buried underground.

Most, but not all, of the mining companies that had abandoned sites on Navajo land had gone out of business, just in time to stick the federal government with the cleanup bills, General Electric, however, had been billed $44 million for its role in the Northeast Church Rock Mine near Gallup. The EPA itself had spent $60 million. Chevron paid an undisclosed amount to clean up the Mariano Lake Mine in New Mexico. "The government can't afford it; that's a big reason why it hasn't stepped in and done more," said Bob Darr, a spokesman for the Department of Energy. "The contamination problem is vast" (MacMillan 2012). Cleaning all the mines would probably cost hundreds of millions of dollars, according to Clancy Tenley, a senior EPA official who oversees its "uranium legacy program."

Many "hot spots" remained. During the summer of 2010, a Navajo cattle rancher, Larry Gordy, found one such site in his grazing land near Cameron, Arizona, 60 miles east of the Grand Canyon. He called the EPA, which found a radioactivity level that, in two days, would expose a human being (or a sheep) to a level of toxicity usually considered unsafe for a year. Such exposure can lead to malignant tumors and other serious health problems, according to Lee Greer, a biologist at La Sierra University in Riverside, California (MacMillan 2012). "If this level of radioactivity were found in a middle-class suburb, the response would be immediate and aggressive," said Doug Brugge, a public health professor at Tufts University Medical School and an expert on uranium. "The site is remote, but there are obviously people spending time on it. Don't they deserve some concern?" (MacMillan 2012).

Radiation levels at more than 80 times the EPA limit were experienced by Mae Begay's family, as described in the film *The Return of Navajo Boy* (2000). Begay, who had been forced to leave her hogan during a cleanup, expressed a strong desire to return home even as Musante admitted that the land "can never go back to what it was before" ("Cleaning Up Uranium" 2011). In 2011, according to the Indian Country Today Media Network, "The Navajo Nation is already dealing with contamination from previous uranium mines and its attendant high rates of cancer, heart disease, and birth defects. Cleanup efforts are taking years, and the U.-S. Environmental Protection Agency (EPA) is evaluating more than 500 sites in the western part of the Navajo Nation."

Uranium tailings cleanup has its own perils. Parts of the Navajo reservation have been so polluted by uranium mining that families whose members have lived there for hundreds of years were being forced to leave their homes and lands so that contaminated soil could be removed. The mines and mills have been closed, but the waste piles endure, polluting water when it rains and creating deadly dust during long, windy, dry periods. Federal environmental officials have been stunned at the extent of uranium contamination.

Aa previously unknown number of previously unrecorded mine sites continue to surface, in many cases after unwary people have fallen into abandoned shafts. "It is shocking—it's all over the reservation," said Jared Blumenfeld, the EPA's regional administrator for the Pacific Southwest. "I think everyone, even the Navajos themselves, have been shocked about the number of mines that were both active and abandoned".

3.12 "Nobody Told Us It Was Unsafe"

In 2014, Bertha Nez, who lived near Church Rock, was facing the prospect of having to abandon her ancestral home at age 67. She had spent months in Gallup motel rooms as radioactive soil was removed from her community. "This is where we're used to being, traditionally, culturally," she said. "Nobody told us it was unsafe. Nobody warned us we would be living all this time with this risk." An enormous pile of uranium mine waste near Nez's home may require 8 years of cleanup. Relocations are voluntary, but many Navajos are skeptical about whether radioactivity can be restored to safe levels around their homes. "Our umbilical cords are buried here, our children's umbilical cords are buried here. It's like a homing device," said Tony Hood, 64, who worked in the mines. "This is our connection to Mother Earth. We were born here. We will come back here eventually".

Between 2008 and 2012, U.S. federal agencies spent $100 million on the cleanup, according to the EPA. An additional $17 million was spent by energy companies responsible for some of the tailings. Waste removal is encumbered by bureaucracy. An application to remove tailings requires paperwork with the NRC, which conducts environmental reviews. Required public hearings also can stall actual removal of toxic waste for several years. "That time frame seems unreasonably long for tribal members, who said that spending so long living away from the reservation has been difficult".

In a decision handed down on December 12, 2013, Allan L. Gropper, a U.-S. bankruptcy judge, said that Anadarko and Kerr-McGee were liable for damages between $4.1 billion and $5.1 billion to a large group of plaintiffs that include "the United States, 22 states, four environmental response trusts and a trust for the benefit of certain tort plaintiffs". The Navajo Nation was scheduled to receive about $1 billion of the settlement to clean up 49 mining and milling sites. As large as the settlement seemed, it left untouched about 90% of the more than 500 abandoned radioactive sites that affect the Navajo.

Navajo Nation President Ben Shelly said that "any funds resulting from this lawsuit are welcomed and long overdue." The Navajo Nation's abandoned uranium mines are part of roughly 2000 sites across the United States that have been "contaminated with various toxic compounds, radioactive waste and other chemicals that are compromising health for numerous communities. The plaintiffs are joined under a litigation trust, which was seeking $25 billion to clean up the U.S. sites".

The Navajos said that, even lacking payment, legal validation had some value: "While we recognize the uncertainties of the appeal process and the long road that may be ahead of us, this is still a day of celebration for the Navajo Nation. A federal judge has issued a ruling that could result in over a billion dollars being made available for cleaning up some of the uranium contamination from past uranium mining and processing on the Navajo Nation".

3.13 The Laguna Pueblo and Anaconda's Jackpile Uranium Mine

Until 1982, the Laguna Pueblo (50 miles east of the Navajo Nation in New Mexico) was the site of the world's largest uranium strip mine. The 7000-acre Anaconda Jackpile mine opened in 1952 and ceased operation when all of the ore that was available at market prices had been mined. Until it closed, the mine made the Laguna Pueblo's people affluent by most reservations' standards. The price of this affluence, in many cases, however, was slow, agonizing death by radiation-caused cancers. When it closed, Jackpile was employing about 800 people. When miners were diagnosed with lung cancer in unusually large numbers, the mine's supporters said what Navajo miners also had been told: they had smoked too many cigarettes. John Redhouse of the Southwest Indigenous Uranium Forum disagreed. "The social costs and health impacts outweigh any jobs and money that goes to Laguna," he said. "Whatever apparent benefits accrue do not necessarily go to the communities but to the multinational energy companies" (Knight 2013).

Decades into the Jackpile mine's active life, some people at Laguna knew they were living on borrowed time. One was the internationally known writer Paula Gunn Allen. Barbara Alice Mann, an honors professor at the University of Toledo and a

3.13 The Laguna Pueblo and Anaconda's Jackpile Uranium Mine 55

friend of Allen's, said that "Paula Gunn Allen was wiped out by the same cancer effects, and even before she was diagnosed, she told me that she knew she would die of cancer, simply because of where she grew up and where her family still lived" (personal communication with Mann, July 25, 2013).

The agricultural valley that had sustained the Laguna Pueblo for centuries before the advent of the mine also was ruined. Before 1952, Rio Paguate had provided water for a valley filled with farms and ranches. Winona LaDuke, writing in *Native Americas*, described the post-mine landscape: "Rio Paguate now runs through the remnants of the strip mine, emerging on the other side a fluorescent green in color" (LaDuke 1992, 58). By 1973, at the latest, the EPA had known that the mine was leaching radiation into the Laguna Pueblo's water supply. By 1975, the EPA had issued reports describing large-scale groundwater contamination. The Laguna tribal center, the community center, and many reservation houses also were dangerously polluted. The problem was compounded when Anaconda used low-grade uranium ore to surface many roads on the reservation.

By 2013, the Jackpile mine was being cleaned up by the Laguna tribe at a cost of $48 million, paid by ARCO. Many Laguna said that even though the "remediation" was supposed to restore the site to its previous condition, uranium had leached into soil and water even before cleaning had been finished, making total cleansing impossible. "Two tributaries near the mine and Rio San Jose have already tested positive for radiation contamination," said Manual Pino of the Laguna-Acoma Coalition for a Safe Environment. "It's one of the best kept secrets of the United States" (Knight 2013).

New uranium mining projects would pose a threat to water supplies on Navajo land, said Kathleen Tsosie, secretary of the Eastern Navajo Diné Against Uranium Mining. Tsosie said that in her home town, Crownpoint, New Mexico, Hydro Resources Inc. has plans "to leach uranium from the groundwater in three places in the Northwestern part of the state. The underground leach mining process is different from traditional open-pit mines since it occurs in the groundwater itself when chemicals are injected into the aquifer to dissolve the ore and is then pumped out. 'How can this not possibly threaten our water supply,' said Tsosie. 'And many of our sacred sites are near these wells'" (Knight 2013). Several Navajo groups petitioned the NRC to revoke the company's mining permit.

"They said the mine would make us rich but I'm still poor and almost everyone around me is dying of cancer and strange diseases," said Dorothy Purley, who was dying of lymphoma after having lived about 1000 yards from the mine. In addition to living close to the mine, Purley also had worked 8 years for Anaconda Jackpile for 8 years, but was not told that radiation placed her life at risk from radiation or that precautions were required to avoid the deadly effects of radiation exposure (Knight 2013). In the area where Purley lived, about 50 miners had died from cancers or related diseases. Another 20 people who lived downwind of the mine but did not work there also had died by 2013, she said.

At the mine itself, she said, lives were in jeopardy every working day. Purley said that in one instance,

> When the crusher got stuck we were down at the bottom using hammers to break the rocks down so the crusher could start going back up into processing, and I helped haul the ore from P9 where they were mining underground [where] the richest ore that was coming out. We hauled back to the ore cars to have it transported. We came home, went to bed, never was a word said that there was contamination [because] of radiation. [I] just realized when I came to deal with this cancer, I found out. This cancer has really taught me a lot … and gotten me to a lot of places and I've realized what cancer really means. ("Uranium Mining" 1996)

They loaded ore into open cars owned by the Santa Fe Railroad, while radioactive dust swirled from the open cars and blew onto everything on its way out of town. "No one tested the tracks for radioactive contamination," she said."The mine's crusher was on the east side of town, as an easterly wind blew yellow powder all over the area" ("Uranium Mining" 1996).

3.14 A Child Nearly Burns to Death

Purley was one of very few women who was hired at the mine. She worked at several jobs, such as receptionist and security guard, often near the mines. "They used to blast sometimes four times a day," she recalled. "When there was no wind, that sulfur that they put in the blasting powder would just kind of sit over the village and sometimes we had to cover our food. We … Indians dried our meat and some of our vegetables out in the sun; we never realized how much contamination there was in the air 'til we realized cancer was the main thing killing our people here in Paguate" (Uranium Mining 1996).

She miscarried three times, as the repeated loss of children broke her heart. Once, before she miscarried, surgery was required after she almost bled to death. Purley knew several women who did not miscarry but gave birth to boys or girls who were deformed. Purley watched a cat as she gave birth to kittens who had no tails. Many children born to women who lived near the miner grew up mentally impaired. All suffered severe allergies of types that had been very rare or unknown before the mine's yellow powder became a fact of every-day life.

Purley suffered from asthma and bronchitis before she was diagnosed with breast cancer. People did not complain because the paychecks put bread and butter on the table. "Money," said Purley, "is the maker of all evil" ("Uranium Mining" 1996).

3.15 Decades of Cleanup Plans

Uranium mining on the Navajo Nation lasted until 1986, when the last of more than 500 mines closed.

A few years after that, following a time-consuming appropriation process by the U.S. Congress, a *very* long cleanup began. By 2020, The EPA continued to solicit bids for various projects. It also released progress reports, eight of them by 2020, such as:

> (April 15, 2020)—Today, U.S. Environmental Protection Agency (EPA) Administrator Andrew Wheeler announced the eighth update to the Administrator's Emphasis List of Superfund Sites Targeted for Immediate, Intense Action.... In this latest update, the Abandoned Uranium Mines contamination in the Navajo Nation was *added* to the list. Thirty-four years and the EPA had the Navajo mines on its priority list (EPA Releases 2020).

In dulcet bureaucratic tones, the EPA spoke of a ten-year plan: to "finalize the *Federal Actions to Address Impacts of Uranium Contamination on the Navajo Nation Ten-Year Plan*." EPA intends to use placement on the Administrator's Emphasis List to focus attention on completion of development and finalization of the Ten-Year Plan. This plan will build on the previous plans, make adjustments based on lessons learned, and identify those next steps necessary to address the human and environmental risks associated with uranium contamination.

"Addressing the legacy of uranium mining continues to be a national effort involving multiple federal agencies, including EPA, collaborating closely with the Navajo Nation government and the respective Navajo Nation agencies," according to a press release on the Internet dated April 15, 2020.

"[The] EPA maintains a strong partnership with the Navajo Nation, and since 1994, the Superfund Program has provided technical assistance and funding to assess potentially contaminated sites and develop a response" (EPA Releases 2020). A strong partnership with the Navajo Nation! Addressing the legacy of uranium mining! A ten-year plan! Identify those next steps necessary! At this rate, the uranium may outlast its half-life by the time those 500 mines are, as they say in EPAland, remediated. After that, the EPA just may turn to the tailings piles that still foul the atmosphere in Navajos' towns and villages and, after that, the Jackpile mine at Laguna Pueblo. And there is more…in Canada, whose Native peoples have their own troubles with the yellow rock.

Wow! If only cleaning up uranium Superfund sites was as easy as writing press releases about it.

3.16 Dene Decimated by Uranium Mining and the "Money Rock"

At the dawn of the nuclear age, Paul Baton and more than 30 other Dene hunters and trappers who had been recruited to mine uranium called it the "money rock." Paid $3 a day by their employers, the Dene hauled burlap sacks of the grimy ore from one of the world's first uranium mines at Port Radium across the Northwest Territories to Fort McMurray. While mining, "many Dene slept on the ore, ate fish from water contaminated by radioactive tailings, and breathed radioactive dust while on the barges, docks and portages. More than a dozen men carried sacks of ore weighing more than 45 kilograms for 12 hours a day, six days a week, four months a year.... Children played with the dusty ore on river docks and portage landings, and their women sewed tents from used uranium sacks" (Nikiforuk 1998, A-1).

Since then, according to an account by Andrew Nikiforuk in the *Calgary Herald*, at least 14 Dene who had worked in the mines between 1942 and 1960 died of lung, colon, and kidney cancers, according to documents obtained through the Northwest Territories Cancer Registry. The Port Radium mine supplied the uranium to fuel some of the first atomic bombs. Within half a century, uranium mining in northern Canada had left behind more than 120 million tons of radioactive waste, enough to cover the Trans-Canada Highway across Canada two meters deep. By 2000, production of uranium waste from Saskatchewan alone occurred at the rate of over one million tons annually (LaDuke 2001).

Dene Cindy Gilday said that uranium mining there that began during the 1940s devastated her hometown of Deline, near Great Bear Lake. Beginning at that time, young Dene men were hired "to carry uranium in sacks from the mines onto barges. The men had no knowledge of the toxic qualities of their loads.... It was the first time people at Great Bear Lake started to die of lung, bone, stomach, brain and skin cancer" (Knight 2013). Since 1975, following a 30-year latency period after uranium mining began in the area, hospitalizations for cancer, birth defects, and circulatory illnesses in northern Saskatchewan increased between 123 and 600%. In other areas impacted by uranium mining, cancers and birth defects increased, in some cases, to as much as eight times the Canadian national average.

"Before the mine, you never heard of cancer," said Baton. "Now, lots of people have died of cancer" (Nikiforuk 1998, A-1). Declassified documents have revealed that the U.S. government, which bought the uranium, and the Canadian federal government, at the time the world's largest supplier of uranium outside the United States, withheld health and safety information from the native miners and their families.

While many of the Dene blame uranium mining and its waste products for their increased cancer rates, some Canadian officials compiled statistics indicating only marginal increased mortality from uranium exposure. André Corriveau, chief medical officer of health for the Northwest Territories, noted that high cancer rates among the Dene do not differ significantly from the overall territorial profile. He said that the death rate was skewed upward by high rates of smoking. The Dene, in the meantime, maintain that the fact that almost half the workers in the Port Radium mine (14 of 30) died of lung cancer cannot be explained by smoking alone.

3.17 "The Incurable Disease"

Until his death in 1940, Louis Ayah, one of the North's great aboriginal spiritual leaders, repeatedly warned his people that the waters in Great Bear Lake would turn a foul, dirty yellow. According to "Grandfather," the yellow poison would flow toward the village, recalled Madelaine Bayha, one of a dozen scarfed and skirted "the incurable disease," "the incurable disease," in the village (Nikiforuk 1998, A-1).

The first Dene to die of cancer, which elders still call "the incurable disease," was Old Man Ferdinand in 1960. He had worked at the mine site as a logger, guide, and stevedore for nearly a decade. "It was Christmastime and he wanted to shake hands with all the people as they came back from hunting," recalled Rene Fumoleau, then an Oblate missionary working in Deline. After saying good-bye to the last family that came in, Ferdinand declared, "'Well, I guess I shook hands with everyone now,' and he died three hours later" (Nikiforuk 1998, A-1).

According to Nikiforuk's account, others died during the next decade. Victor Dolphus's arm came off when he tried to start an outboard motor. Joe Kenny, a boat pilot, died of colon cancer. His son, Napoleon, a deck hand, died of stomach cancer. The premature death of so many men has not only left many widows but also interrupted the transmission of culture. "In Dene society it is the grandfather who passes on the traditions and now there are too many men with no uncles, fathers or grandfathers to advise them," said Cindy Gilday, Joe Kenny's daughter and chair of Deline Uranium Committee (Nikiforuk 1998, A-1).

"It's the most vicious example of cultural genocide I have ever seen," Gilday said. "And it's in my own home." According to Nikiforuk, "Watching a uranium miner die of a radioactive damaged lung is a job only for the brave." He described Al King, an 82-year-old retired member of the United Steelworkers union in Vancouver, British Columbia, who has held the hands of the dying. King described one retired Port Radium miner whose chest lesions were so large that they had spread to his femur and exploded it. "They couldn't pump enough morphine into him to keep him from screaming before he died," said King (Nikiforuk 1998, A-1).

In exchange for their labors in the uranium mines, each Dene received a few sacks of flour, lard, and baking powder every few weeks. "Nobody knew what was going on," recalled Isadore Yukon, who hauled uranium ore for three summers in a row during the 1940s. "Keeping the mine going full blast was the important thing" (Nikiforuk 1998, A-1).

> The Dene town of Deline was described by one of its residents as practically a village of widows. Most of the men who worked as laborers have died of some form of cancer. The widows, who are traditional women were left to raise their families with no breadwinners, supporters. They were left to depend on welfare and other young men for their traditional food source. This village of young men, are the first generation of men in the history of Dene on this lake, to grow up without guidance from their grandfathers, fathers and uncles. This cultural, economic, spiritual, emotional deprivation impact on the community is a threat to the survival of the one and only tribe on Great Bear Lake. (Gilday n.d.)

3.18 The Conflict Continues

During and after 2019, many Dene continued to protest uranium mining on their lands, to the point of setting up blockades against industrial traffic aiming to open one of the largest such mines in North America. Organized as the Northern Trappers Alliance, the Dene also opposed exploitation of tar sands. Press reports said that the blockade, established November 22, 2014, was "forcibly dismantled" by officers of the Royal Canadian Mounted Police December 1. Eighty days, later wrote Michael Toledano in *Vice*, "The trappers remain [ed] camped on the side of the highway in weather that has routinely dipped below -40°C. They are constructing a permanent cabin on the site that will be a meeting place for Dene people and northern land defenders."

> "We want industry to get the hell out of here and stop this killing," said Don Montgrand, who has been at the encampment since day one and was named as one of its leaders on the police injunction. "We want this industry to get the hell out before we lose any more people here. We lose kids, adults, teenagers." "They're willing to stay as long as it takes to get the point across that any of this kind of development is not going to be welcomed," said Candyce Paul, the alliance's spokesperson and a member of the anti-nuclear Committee for the Future Generations. "It's indefinite." "We don't want to become a sacrifice zone. That's where we see ourselves heading".

The trappers have been reacting to a rise in cancer rates among the Dene, as well as industrial development that drives away game on which they depend for income, food, and sustenance of traditional culture. In its place, Fission Uranium Corp. wants to develop a high-grade uranium mine that could become one of the largest on Earth, extending a 150-mile long "integrated uranium corridor," which already hosts "the largest high-grade uranium mines and mills in the world, with their own stockpiles of radioactive tailings and a decades-long history of radioactive spills".

All of this is being built upon the ruins of earlier mines that killed many Dene in earlier generations, "abandoned and decommissioned uranium mines already host millions of tons of radioactive dust (also known as tailings) that must be isolated from the surrounding environment for millennia, while no cleanup plans exist for the legacy of severe and widespread watershed contamination that is synonymous with Uranium City, Saskatchewan". More than 85% of the people in this area (northern Saskatchewan) are aboriginal. They are fighting exploratory mining that could increase in scale in coming years, a new neighbor to the open-pit tar sands mines further south which have turned indigenous lands into "moonscapes."

"When they spew the pollution, it affects our water, lakes, fish—any kind of species. Our traditional life destroyed with these oil mines around us," said Kenneth, one of the trappers. "We're in the middle of these oil mines and the government's still not listening." "We know our water isn't as good as it used to be," said Paul. "You see more fish with lesions".

3.19 Plans to Blast an Alaskan Harbor with Nuclear Bombs

At times, nuclear weapons have been proposed as construction devices in Indian Country. This is not a parody of Dr. Strangelove. During the 1950s, seeking "peaceful" uses for the weapons, the U.S. Atomic Energy Commission (AEC) actually proposed dropping five atomic bombs at Cape Thompson on the North Slope of Alaska to blast a harbor. Project Chariot, as it was called, was part of the agrarian-sounding Project Ploughshare, from the biblical concept of peacemaking (beating swords into ploughshares), a piece of bureaucratic poetry supposedly invoking civilian uses for nuclear bombs.

The plans were precise—five bombs with 100 times the explosive power of the first atomic bomb that had targeted Hiroshima. The AEC and other agencies recruited Alaskan political leaders and newspaper publishers, but residents in the Inupait village of Point Hope, about 450 people living 30 miles from the intended target site, rallied against the harbor project as a horrid ecological insanity. "Project Chariot Still On," read a banner headline at the top of the front page of the *Tundra Times*.

The Inupiat fought off the project, but radioactive traces that had been used in preliminary tests later were found at the site. The AEC planted radioactive materials in 15,000 pounds of soil obtained from a Nevada nuclear test site (on Western Shoshone land) around Point Hope during 1962. The Native people of the Point Hope area were never informed of other experiments on their homeland. The U.-S. government tested the effects of radioactive bioaccumulation on caribou, lichen, and humans in the area. One such project was made public in 1992 following a Freedom of Information Act request. The residents demanded and won cleanup by the U.S. Department of Energy more than 30 years after they had unknowingly been used as test subjects with live radioactivity (LaDuke and Cruz 2013, 43).

3.20 The Point Hope Eskimos: An Atomic Harbor and a Nuclear Dump as a Neighbor

At times, nuclear weapons have been proposed as construction devices in Indian Country. This is not a parody of Dr. Strangelove. During 1958, seeking "peaceful" uses for the weapons, the Atomic Energy Commission (AEC) seriously, officially, proposed dropping five atomic bombs at Cape Thompson on the North Slope of Alaska to blast out a harbor. Project Chariot, as it was called, was part of the agrarian-sounding Project Ploughshare, from the biblical concept of peacemaking, beating swords into ploughshares, a piece of bureaucratic poetry invoking civilian uses for nuclear bombs. Project Chariot's authors had forgotten, disregarded, or did not know that they were proposing bombing the homeland of the Point Hope Eskimos, on the North Slope of Alaska. The plan called for underground explosions that would force land to sink into the sea.

The plans were precise—five bombs with 100 times the explosive power of the first atomic bomb that had obliterated and radiated Hiroshima and Nagasaki. The AEC and other agencies recruited Alaskan political leaders and newspaper publishers, but residents in the Inupait village of Point Hope, about 450 people living 30 miles from the intended target site, rallied against the harbor project as a horrid ecological insanity. "Project Chariot Still On," read a banner headline at the top of the front page of the *Tundra Times*. They were wrong—but sanity took some time to cultivate.

3.21 Nuclear Boosterism

When word reached people on Point Hope that their homeland was being suggested as a sacrifice area for nuclear boosterism, who have inhabited their homeland for more than 2000 years, they were outraged, especially after the U.S. Bureau of Land Management approved the AEC's request for a test site.

The plan to blast a harbor at Point Hope had been championed by Edward Teller (1908–2003), who was popularly known as the "father" of the hydrogen bomb, although he disdained the moniker. Notwithstanding his preference in nicknames, Teller, who also was known for his explosive personality, was an advocate of using nuclear energy to address many technological problems, of which the idea of opening a harbor near Point Hope was one. Teller toured Alaska promoting the idea as the territory was preparing for statehood, and at one point rallied newspaper editors and civic boosters behind the idea. Given his recklessness with nuclear bombs, Teller has long been considered an inspiration for the main character in Stanley Kubrick's "Dr. Strangelove," (1964), a parody of atomic insanity. This proposal, however, was no Hollywood movie.

The Plowshare name was attributed to The *Bible's Isaiah 2:3-5*, a passage focused on repurposing the weapons of war for the tools of peace:

> They will beat their swords into plowshares and their spears into pruning hooks. Nation will not take up sword against nation, nor will they train for war anymore.

Teller believed that nuclear blasts could be contained, and that a radiation-free bomb was on the way. He loved to imagine that nuclear explosions were so easily controlled that the Chariot team could "dig a harbor in the shape of a polar bear, if desired" (Mead 2012). He called the use of atomic bombs as excavation devices "the nuclear shovel." "Among other things the Point Hope people were told that the fish in and around the Pacific Proving Grounds were not made radioactive by nuclear weapons tests and [there would not be]... any danger to anyone if the fish were utilized; that the effects of nuclear weapons testing never injured any people, anywhere..." (Mead 2012).

Operation Plowshare had plans for much more than Point Hope. According to one retrospective (Mead 2012):

> Plowshare's vision was as enormous as its tools. One proposal, called Project Carryall, would have used nuclear bombs to cut a new 11,000 foot long railway pass through the Bristol Mountains in California. According to a 1964 feasibility study, Carryall would have required 22 bombs ranging between 20 and 200 kilotons to gouge out the deepest parts of the pass, which would have measured about 350 feet deep. Parts of the pass about 100 feet deep would have been dug with conventional non-nuclear bombs. The nukes were to be saved, as with the Alaska project, for the heavy lifting.

The Plowshare project produced 27 nuclear test explosions, most of them at a federal government nuclear test site in Nevada.

> Project Sedan...was the second Plowshare experiment to be carried out. It basically was a test to see how big of a hole a nuclear bomb could make. It proved to be a *really big* hole. The 104-kiloton device moved 12 million tons of earth, producing the largest man-made crater in the country, which measured 1280 feet wide and 320 feet deep. Shot in Nevada, Sedan spewed fallout over Iowa, Illinois, Nebraska, and South Dakota, and contaminated more Americans than any other nuclear test (Mead 2012).

Operation Plowshare also played a role in the development of natural-gas drilling technology. Having tried dynamite, machine guns, bazookas, and napalm to extract the gas from unforgiving (the oil-industry jargon is "tight") underground rock formations, Operation Plowshares as called upon to pulverize it. "In 1967," wrote Lawrence Wright in *The New Yorker* (2018, 46), "the Atomic Energy Commission, working with Lawrence Livermore Laboratory and the El Paso Natural Gas Company, exploded a 29-kiloton nuclear bomb, dubbed Gasbuggy, 4,000 feet below the surface, near Farmington, New Mexico, [at the northeastern edge the Navajo Nation]." The explosions proved successful at liberating the gas from the pulverized rubble, but one side effect that the project's engineers should have anticipated made it unmarketable. The gas was radioactive.

3.22 An "Overture to the New Era"?

The proponents of Project Chariot seem to have forgotten that the Point Hope area was populated. An editorial in the July 24, 1960, *Fairbanks News-Miner* said, "We think the holding of a huge nuclear blast in Alaska would be a fitting overture to the new era which is opening for our state" (Mead 2012). A few conservationists joined the Eskimos in opposition, which grew when news of the proposal reached a broader ecologically oriented community in the "Lower 48," including the Sierra Club, the Wilderness Society, and the Committee on Nuclear Information, led by Barry Commoner. The growing wave of popular revulsion against the idea provoked the AEC to suspend Project Chariot in 1962, although it was never fully canceled.

Aside from its potential ecological damage and radiation exposure to the Eskimos, the atomic harbor project had other problems, once the boosters' gloss wore off. One major obstacle was the fact that the harbor had absolutely no commercial potential. Nuclear boosters could not leave the site alone, however. They turned Project Chariot into a study of how nuclear fallout could affect the indigenous communities of Point Hope, Noatak, and Kivalina. "to measure the size of bomb necessary to render a population dependent" after local food sources have become too dangerous to eat due to extreme levels of radiation" (Davis 1973, 143–151).

3.23 Radiation Tests' Effects on Eskimos

After the harbor project was suspended, the AEC conducted environmental studies which suggested that radioactive contamination from the blasts that it had proposed would have adversely affected local people as it radiated the animals they hunted. The AEC studies indicated that even without local atomic blasts, radiation from other blasts' fallout in distant locations was moving with unexpected rapidity up the Eskimo food chain from lichen which was consumed by caribou, which then were eaten by the native peoples. The Eskimos paid for these tests with abnormally high rates of cancer for at least 30 years afterwards (Vandegraft 1993).

The Point Hope site also was radioactively contaminated during an experiment that gauged effects of fallout on plants growing on the tundra that later was spread by rainfall. Soil from 1962 nuclear test explosion in Nevada was used in tests at the Point Hope site and then buried, where it remained for three decades until it was re-discovered and removed.

Even though the Inupiat fought off the project, radioactive traces that had been used in preliminary tests later were found at the site. The AEC planted radioactive materials in 15,000 pounds of soil obtained from a Nevada nuclear test site (on Western Shoshone land) around Point Hope during 1962. The Native people of the Point Hope area were never informed of other experiments on their area. The U.S. government tested the effects of radioactive bioaccumulation on caribou, lichen, and humans in the area. One such project was made public in 1992 following a Freedom of Information Act request. The residents demanded and won cleanup funds from the U.S. Department of Energy more than 30 years after they had unknowingly been used as test subjects with live radioactivity (LaDuke and Cruz 2013, 43).

The 43 pounds of radioactive soil that was stored near Point Hope came from within a mile of ground zero of a nuclear blast in Nevada; it contained strontium-85 and cesium-137 (Badger 1992, B-5). The strontium typically would have lost all its radioactivity years before its deposit at Point Hope, and the cesium would still have had about half its radioactivity after 30 years, according to government officials. The purpose of the experiment, according to the A.E.C., was to test the toxicity of radiation in an arctic environment. The dump experiment was carried out by the U.S. Geological Survey under license from the A.E.C.

The Inuit did not learn of the nuclear dump until Dan O'Neill, a researcher at the University of Alaska, made public documents he had found as he researched a book on the aborted plan to create a harbor in the Alaskan coast with nuclear weapons. O'Neill, using the Freedom of Information Act, learned that the nuclear waste had been stored in the area as part of Project Chariot, which was declassified in 1981.

3.24 Cancer Rates Rise

For many years, the Inuit in the area had suffered cancer rates that far exceeded national averages. The government acknowledged that soil in the area contains "trace amounts" of radiation, but denied that its experimental nuclear dump had caused the Inuits' increased cancer rate. According to O'Neill, the nuclear dump was clearly illegal, and contained "a thousand times…the allowable standard for this kind of nuclear burial" (Grinde and Johansen 1995, 238–239).

The nuclear waste that was buried near Point Hope had remained unmarked for 30 years, during which time hunters crossed it to pursue game, and caribou migrated through it. Not until September 1992 did the U.S. federal government admit that it had buried 15,000 tons of radioactive soil at Cape Thompson, 25 miles from Point Hope, on the Chukchi Sea in northwestern Alaska. Until the dumps were disclosed during the late 1990s, the Inuit in Port Hope had no clue as to why the incidence of cancer in their village had jumped to 578 per 100,000 within two generations. Some doctors blamed the rise in cancer rates on smoking by the Inuit. In 1997, Dr. Bowerman, chief medical officer of the Borough of Barrow, Alaska, published findings linking the increase in cancer incidence to the burial of nuclear waste near Port Hope (Colomeda 1998).

"I can't tell you how angry I am that they considered our home to be nothing but a big wasteland," said Jeslie Kaleak, mayor of the North Slope Borough, which governs eight Arctic villages, including Point Hope. "They didn't give a damn about the people who live up here." When Senator Frank Murkowski, Republican of Alaska, visited the village, an elderly woman threw herself at him and shouted, "You have poisoned our land!" (Egan 1992, A-26).

Energy Department spokesman Tom Gerusky acknowledged that the Geological Survey erred in burying the waste but said a person standing on the mound for a year would be exposed to only a small fraction of radiation received on a single cross-country jet flight. (Badger 1992, B-5) A series of studies in the 1980s by the federal Centers for Disease Control and the Indian Health Service concluded that while radiation from Soviet tests was detected in a number of Alaska villages during the 1960s, the bulk of cancers in the region involve lung and cervical cancer, types that are not generally associated with exposure to radiation.

3.25 The Prairie Island Indian Community and Nuclear Waste

The 880-member Prairie Island Indian community in southeastern Minnesota was told that a nuclear waste dump next to their reservation would be temporary, but Xcel Energy, which owned a power plant nearby, had come to stay. Although the indigenous community had been told that the dump would close in the 1990s, in 2012 the tribe was told that Xcel wanted to extend the life of its "dry cask" storage facility (which warehouses spent fuel rods) another 40 years. About 200 Native people live on the reservation, and as many as 8000 people a day visit its Treasure Island Resort and Casino (Shaffer 2012).

"Trust is already an issue for the Prairie Island community," reported the Indian Country Today Media Network, "given incidents like the temporary shutdown of one of the facility's reactors on August 14 [2012], when two diesel generators broke down during a routine monthly maintenance test.... Failure of the generators, which are in place to help the plant operate during power outages, did not inspire confidence from the tribe even though Xcel assured tribal members that the outage posed no danger.... These assurances ... are reminders of the fact that the 40-year old Prairie Island Nuclear Generating Plant operating a half-mile from our homes relies on aging technology," the Prairie Island Indian Community said. "To have not just one, but both of the back-up diesel generators fail is very troubling. A failure of the back-up diesel generators can affect all other safety features that rely on the electricity that they generate. The failure of both of them during a routine monthly test is simply not acceptable.... Our frustrations continue to grow over the federal government's failure to live up to its responsibility to adequately address the nuclear waste issue and as more of that toxic waste builds up next to our community" ("Prairie Island" 2012).

3.26 The Western Shoshone: "The Most-Bombed Nation of Earth"

Since 1985, many Western Shoshone people, who call themselves "the most-bombed nation on Earth," have joined with environmentalists to resist nuclear weapons testing in Nevada, northwest of Las Vegas, trespassing on the world's only remaining nuclear weapons proving ground. The test site occupies land that President Harry Truman confiscated from the Western Shoshone for national security purposes during 1951, forcibly relocating 100 Native American families.

Traditional Shoshone medicine man Corbin Harney (1920–2007) "came out from behind the bush," as he says, in 1985, initiating protests of the U.S. government's nuclear weapons testing (Harney 1995). (Harney died on July 10, 2007, at age 87, of non-Hodgkin's lymphoma, a cancer that had metastasized into his bones.) He also had taken hundreds of people (Indian and non-Indian) into sweat lodges to pray against testing. Corbin and other Western Shoshone leaders repeatedly crossed a gated cattle guard onto the 1350-square-mile test site. On occasion, they were joined by large groups of people marching in support of a ban on nuclear testing. The marches, which have included several thousand people, sometimes are held on Mother's Day (for Mother Earth). At some of these marches, hundreds of people have been arrested.

Harney cited evidence that the underground tests leak radiation aboveground. "Down-winders," people who live downwind from nuclear weapons test sites, especially in southern Utah, had, according to Harney, "observed extremely high numbers of cancers, leukemia, and other physical deformities in their populations." Harney compared these reports to similar findings among citizens of Kazakhstan (in the former Soviet Union) who lived downwind of a nearby test site. They have reported similarly high levels of cancer, leukemia, and birth deformities, among other illnesses. Protests later closed the test site in Kazakhstan. During 1993, when Harney visited Kazakhstan, he learned that local water had been irreversibly contaminated with dangerous levels of radiation. "You can't drink a glass of water there anymore," he said. "All they had to drink was vodka, cartons of juice, and bottled water imported from Europe" (Harney 1995).

3.27 More Tests on Indian Land

Since 1951, about 700 nuclear tests for Great Britain and the United States have since been conducted at the Nevada Test Site on lands belonging to the Western Shoshone, according to the Treaty of Ruby Valley, ratified in 1863. Until 1963, the tests were conducted aboveground, a form of testing that was banned by international treaties during the 1960s; in the 1970s, tests also were limited to a 150-kiloton explosive yield.

The EPA has documented cumulative deposits of plutonium in soil samples more than 100 miles north of the test area. "We are the most-bombed nation in the world," said William Rosse Sr., a Western Shoshone elder. "We've had more than our share of radiation, and now they want to dump more nuclear waste on our land at Yucca Mountain" ("Havasupai Fight" 1992). "We have always been a free tribe, never have we been conquered and never have we sold land to the U.S.," said John Wells of the Western Shoshone National Council. "My people are still suffering from the 828 underground nuclear weapons tests and 105 above-ground nuclear weapons tests at the Nevada Test Site" (Hodge 2001).

3.28 Washington State's Yakamas and Hanford's Radioactive Legacy

The Hanford Nuclear Reservation in eastern Washington State (within the treaty boundary of the Yakama Nation) has become a storage site for roughly 53 million gallons of high-level chemical and radioactive wastes that released irradiated water 30 miles upstream from the Yakama Reservation between 1945 and 1989. The presence of this porous depository in their midst has exposed some people to levels of radiation similar to that of the Chernobyl nuclear accident in Russia.

Due to this decades-long bath of radiation dumped directly into the river system, oysters caught at the mouth of the Columbia River were so toxic by the early 1960s that when one Hanford employee ate them and returned to work the following day, he set off the plant's radiation alarm. An investigation revealed that the day before, he had eaten a can of oyster stew contaminated with radioactive zinc. The oysters had been harvested in Willapa Bay, along the Pacific Coast in Washington State, 25 miles north of Astoria, Oregon, having been bathed in radioactive waster that had flowed from Hanford, halfway across Washington State, a couple of hundred miles into the Pacific, then north along the coast, at least.

The 560-square-mile Hanford Nuclear Reservation was established during 1943 on lands traditionally used for hunting, fishing, and gathering by the Yakama and Umatilla. The same area is adjacent to the homeland of the Nez Perce. The Hanford facility produced the plutonium used in the first atom bombs dropped on Hiroshima and Nagasaki. Thus, releases of radioactive materials have been contaminating these peoples, as well as the Coeur d'Alene, Spokane, Colville, Kootenai, and Warm

3.28 Washington State's Yakamas and Hanford's Radioactive Legacy

Springs tribes, for many years. In 1986, after disclosure that radiation was secretly released into the air and water from 1944 until January 1971, when the last of Hanford's reactors was closed, Yakama leaders were among the first to call for a thorough study of the danger.

For nearly 30 years, ending in 1971, eight of the nine nuclear reactors at the Hanford complex were cooled by water from the Columbia River. Millions of gallons of water that had been pumped directly through the reactor cores picked up large amounts of nuclear material, making the Columbia downstream the most radioactive large river in the world, according to state and federal authorities (Schneider 1990, A-9). In July 1990, a federal panel said that some infants and children in the 1940s absorbed enough radioactive iodine to destroy their thyroid glands and cause an array of thyroid-related diseases. Although Hanford has stopped producing plutonium, radioactive materials leaking from storage tanks there have continued to contaminate groundwater in the area. The Indigenous Environmental Network reported that "the Hanford site also has eight nuclear reactors that have ... contaminated the Columbia River on which many of the Tribes depend for basic sustenance" ("Uranium/Nuclear" 2001). Unaware of the contamination, indigenous people collected berries near the Columbia, hunted for eels in its tributaries, and fished salmon from its waters.

New York Times reporter Keith Schneider (1990) described Colombia below the Hanford Reservation:

> Dammed in the 1950s below Hanford and developed by industrial companies, the river's water is green and gray now. Salmon runs are much smaller than they were before World War II, and Johnny Jackson, a 59-year-old fisherman ... said some fish he catches were marred by deep, infected welts and growths.... Documents declassified beginning in 1986 said that "radiation spread to the river's bacteria, algae, mussels, fish, birds, and the water used for both irrigation and drinking." (A-9)

Radiation in the river has caused concern at Hanford for decades. In 1954, the situation was reviewed in closed meetings at the Washington, D.C., headquarters of the AEC, which operated the plant. Lewis L. Strauss, the commission's chairman, flew to Hanford in the summer of 1954 and was told that "levels of radioactivity in some fish in the Columbia River, particularly whitefish, were so high that officials were considering closing sport fishing downstream" (Schneider 1990, A-9). Ducks, geese, and crops irrigated with Columbia River water also were said to pose a potential health threat.

The public never was alerted. As Schneider (1990) wrote, in a memorandum prepared on Aug. 19, 1954, Dr. Parker urged the government to keep the problem secret because the radiation levels were still within safety guidelines and closing sport fishing would compel the plant to discuss the issue publicly and compromise secrecy. "The public relations impact would be severe," he wrote. (A-9).

Secrets, like radioactivity, are prone to leak as their sources age—here and elsewhere in Indian Country, where recovery from a deadly radioactive legacy, when and if it comes, has been slow and painful.

3.29 On the Road to Hell, the Most Toxic Lies Go Undisclosed

In Northwest Canada, as late as November 2021, the Dene were still trying to get uranium explorers, miners, and millers off their land. Very unlikely unaware of the pain and death that the "money rock" had inflicted on Dene people, the mining companies kept coming, and when the Native owners of the land warned them away, they *still* kept coming, and staying. The government of the Clearwater River Dene Nation "served notice on the uranium industry and the governments of Canada and Saskatchewan regarding the community's grave concerns about the potential impacts and risks posed by an increasing number of uranium mining and milling projects and exploration activities occurring within its traditional lands" (Clearwater River 2021).

In the worst "tradition" of Euro-American-led governments across the Americas for hundreds of years, the provincial government of Saskatchewan had issued permits for uranium exploration without asking the Dene.

> Old story, the Dene's government said. Our people have suffered enough. "Over the past few years, numerous mining companies have conducted uranium exploration and have proposed extensive development in the heart of CRDN's Traditional Lands. These included proposals for mines by NexGen Energy Ltd. and Fission Uranium Corp. and exploration activities by Denison Mines Corp., Standard Uranium, Purepoint Uranium, Cameco, UEX Uranium, Orano, Azincourt Energy, and other lease-holders' affiliated proponents. The Government of Saskatchewan has granted tenure and issued authorizations for exploratory activities in the absence of meaningful consultation with the CRDN.
>
> CRDN General Manager Walter Hainault stated, "contrary to the United Nations Declaration on the Rights of Indigenous People (UNDRIP) and directives of the Supreme Court of Canada, governments have and continue to illegally permit uranium companies to run roughshod over our People, Traditional Lands and Treaty Rights. As a result, we now have uranium mines being proposed in the Patterson Lake Area—the hub and heartland of our ancient and traditional lands." (Clearwater River 2021).

The statement continued:

The Clear Water Dene has conducted its own in-depth Indigenous rights and knowledge study with elders, trappers, hunters, fishers, youth, and families, who reiterated how often they had been forced off their lands by uranium mining, and how they have been impacted and are being forced out of their traditional family hunting grounds by uranium mining activity; how much they depend on their land for sustenance; why the Dene oppose future mining as incompatible with CRDN's rights, culture, and way of life, and how all of this has been codified in Canada's own laws. ""We are now preparing for [the] upcoming environmental assessment processes, and regulatory hearings. Our People will be front and center and their voices will be heard and all of the issues, concerns and alternatives they bring forward must be addressed" (Clearwater River 2021).

3.30 The Dene Extend Profound Condolences to the Victims in Hiroshima and Nagasaki

When money was at stake, the miners' memories were short, very short, as so often has been the case. The Dene were saying "Not again. Stop! This is our land, and mining of uranium has no place in it. Experience tells us that we must and will struggle to protect our land, people, and culture. Our memories are long and our experiences have been painful. Even after many Dene had died from affliction by cancers, they learned that some of the uranium used in the atomic bombs dropped on Japan as World War II was ending had been mined from Dene land. The Dene people were greatly saddened by this knowledge, so much so that they sent a delegation to Hiroshima with profound condolences.

> The existence of this epic road to hell was unsuspected by the Dene until long after Eldorado stopped mining for radium and uranium in 1960. Beginning in the 1970s, and spiking sharply in the 1980s, many of the men who had handled and carried the ore—and the men who had mined it—began to die from cancer, raising obvious questions about health and safety which soon led in shocking directions. In the 1990s, Gordon Edwards, co-founder of the Canadian Coalition for Nuclear Responsibility (CCNR), told a group of Déline Dene of compelling evidence that the ore that had made so many of them sick had been used to kill vast numbers of innocent people. By the end of decade, Déline was better known as The Village of Widows, the title of a film by Peter Blow, documenting not only the toll on the community but its extraordinary Hiroshima pilgrimage."(Howard 2019).

This news broke on the Canadian public in 1998 after "The Village of Widows" was broadcast on the Canadian Vision Channel and picked up by the Canadian

Broadcasting Corporation (CBC). The CBC quoted Dene sources indicating that they "were never told of the health hazards they faced, even though the government knew ... as early as 1932 that precautions should be taken in handling radioactive materials. Instead, they were told that the work was safe."

Reality was much crueler. For example, Paul Baton, aged 83, [in 2019], used to lift sacks of uranium ore onto boats. "Paul" Baton [actually Peter Baton] was interviewed by scholar and playwright Julie Salverson for her *Lines of Flight, an Atomic Memoir,* published in 2016. Peter's wife, Theresa, told her: "...that when they lived in Port Radium, the women would make tents for their families to sleep in from the sacks that carried the uranium." This informastion had begun in a December 1998 article, "Déline Dene Mining Tragedy," in the *First Nations Drum.* (Howard 2019). The same year, the Déline Dene Band Uranium Committee released a 160-page report, *They Never Told Us These Things.* The biggest lies go unsaid.

The Village of Widows was broadcast on the Canadian Vision channel in 1998; the following June, without mentioning the film, CBC News announced that it "has learned about a dark chapter in Canadian history," a scandal surrounding "the raw materials used to make the atomic bombs that fell on Japan."

The actual mining was done by non-Indians brought in from outside by Eldorado Gold Mines Ltd. *The Dene, who were paid three Canadian dollars a day as "coolies," carried 70-pound bags of radioactive uranium ore 12 hours a day, six days a week.* At that time, the ore itself was worth more than Canadian $70,000 a gram. Cindy Kenney-Gilday, a member of the delegation to Hiroshima and Chair of the Déline Dene Band Uranium Committee, told *First Nations Drum* journalist Ronald B. Barbour that the toll taken on "my own home" is "the most vicious example of cultural genocide I have ever seen" (Howard 2019).

"Kenney-Gilday," Barbour wrote, who has suffered the loss of her father to colon cancer and a brother to stomach cancer, stressed that because "it is the grandfathers in Dene society" who traditionally transmit teachings and world-ways to younger men and boys, "the loss of these men in the community" has left "too many men without guides". The trail to these deaths was paved with more than 1.7 million tons of radioactive waste dumped on Dene land and lakes, making fish and other food inedible. After all of this, the errand fell to the Dene to travel to Hiroshima, where "One of the widows expressed deep sorrow that the material that came out of our land had killed innocent victims in a land that's foreign to us." "Yet the most extraordinary apology in Canadian history," wrote Howard (2019) "was surely that offered by the victims of systematic mistreatment by the Canadian government to the victims of a crime against humanity [that] they unknowingly [had] helped others commit" (Howard 2019). On 6 August 1998, 10 members of the small Sahtúgot'ine Dene community of Déline (Fort Franklin) in the Northwest Territories apologized in Hiroshima for the atomic destruction of that city—and the death of over 200,000 civilians—exactly 53 years earlier by a bomb made in part from uranium from their land. (Howard 2019).

References

American Diagnosis: A co-production of KHN and Just Human Productions. Podcast, with Dr. Celine Gounder, KHN, February 15, 2022. Season 4. https://khn.org/news/article/american-diagnosis-episode-3-uranium-mining-left-navajo-land-and-people-in-need-of-healing/. Last accessed March 17, 2022

Arnold, Carrie. 2014. "Once upon a Mine: The Legacy of Uranium on the Navajo Nation." *Environmental Health Perspectives.* http://ehp.niehs.nih.gov/122-A44. Last accessed Much 18, 2022.

Badger, T.A. "Villagers Learning a Frightening Secret: U.S. Reveals that it has Buried Radioactive Soil Near Alaska Town 30 Years Ago. Residents Fear that Atomic Testing may have Damaged the Food Chain." Los Angeles *Times*, December 20, 1992, B-5.

Calvert, Mary F. "Toxic Legacy of Uranium Mines on Navajo Nation Confronts Interior Nominee Deb Haaland." February 23, 2021. https://pulitzercenter.org/stories/toxic-legacy-uranium-mines-navajo-nation-confronts-interior-nominee-deb-haaland. Last accessed May 1, 2022.

Cleaning Up Uranium in the Navajo Nation. 2011. Indian Country Today Media Network, May 31. Last accessed March 4, 2022.

"Cleanup Planned for Largest Abandoned Uranium Mine on the Navajo Nation". 2011. Environment News Service, September 29. http://www.ens-newswire.com/ens/sep2011/2011-09-29-092.html. Last accessed June 22, 2010.

"Clearwater River Dene Nation Serves Notice on Uranium Industry Regarding Impacts of Uranium Mines and Exploration". Clearwater River Dene Nation. November 10, 2021. https://www.newswire.ca/news-releases/clearwater-river-dene-nation-serves-notice-on-uranium-industry-regarding-impacts-of-uranium-mines-and-exploration-898283328.html. Last accessed May 17, 2022.

Colomeda, Lori A. (Salish Kootenai College, Pablo Montana). "Indigenous Health." Speech. Brisbane, Australia, September 9, 1998. http://www.ldb.org/vl/ai/lori_b98.htm. Last accessed March 15, 2022.

Csordas, T. J. 1999. "Ritual Healing and the Politics of Identity in Contemporary Navajo Society." *American Ethnologist* 26: 3–23.

Davis, Robert. *The Genocide Machine in Canada: The Pacification of the North.* Black Rose Books, 1973, 143-151.

Egan, Timothy. "Eskimos Learn They've Been Living Amid Secret Pits of Radioactive Soil." *New York Times*, December 6, 1992, A-26.

Eichstaedt, Peter. 1994. *If You Poison Us: Uranium and American Indians.* Santa Fe, NM: Red Crane Books.

"EPA Awards Contracts Worth up to $220 million to Three Companies for Cleanup at Navajo Nation Area Abandoned Uranium Mines". February 11, 2021. https://www.epa.gov/newsreleases/epa-awards-contracts-worth-220-million-three-companies-cleanup-navajo-nation-area. Last accessed March 17, 2022.

"EPA Releases Eighth Update to the Administrator's Superfund Emphasis List, Navajo Abandoned Uranium Mines added". U.S. Environmental Protection Agency. April 15, 2020. https://www.epa.gov/newsreleases/epa-releases-eighth-update-administrators-superfund-emphasis-list-navajo-abandoned. Last accessed March 18, 2022.

Gilday, Cindy Kenny. n.d. "A Village of Widows." *Arctic Circle.* http://arcticcircle.uconn.edu/SEEJ/Mining/gilday.html. Last accessed May 20, 2012.

Grinde, Donald A., Jr. and Bruce E. Johansen. *Ecocide of Native America: Environmental Destruction of Indian Lands and Peoples.* Santa Fe, NM: Clear Light, 1995.

Harney, Corbin. 1995. *The Way It Is: One Water, One Air, One Earth.* Nevada City, CA: Blue Dolphin Publishing.

Hodge, Damon. 2001. "Dumpy Meeting: Final Vegas Hearing on Yucca Mountain Plagued by the D.O.E." *Las Vegas Weekly*, September 11. http://www.lasvegasweekly.com/2001_2/09_13/news_upfront7.html. Last accessed August 1, 2001.

Howard, Sean. "Canada's Uranium Highway: Victims and Perpetrators." *Cape Brerton Spectator.* August 7, 2019. https://capebretonspectator.com/2019/08/07/canadas-uranium-dene-bomb/. Last accessed May 17, 2022.

Ivins, Molly K. 1978. "Doctor Charges U.S. Ignored Uranium Spill." *Hutchinson (Kans.) News,* July 27, n.p.

Johansen, Bruce E. 1995. "Victims of Progress: Navajo Miners" [Review of Eichstaedt, *If You Poison Us*]. *Native Americas* 12:3 (Fall): 60–61.

Johansen, Bruce E. 1997. "The High Cost of Uranium in Navajoland." *Akwesasne Notes New Series* 2, no. 2 (Spring): 2010. http://ratical.com/radiation/UraniumInNavLand.html. Last accessed September 7, 2021.

Knight, Danielle. 2013. "Native Americans Denounce Toxic Legacy." TWN: Third World Network, June 14. http://www.twnside.org.sg/title/legacy-cn.htm. Last accessed July 1, 2014.

LaDuke, Winona. 1992. "Indigenous Environmental Perspectives: A North American Primer." *Native Americas* 9, no. 2 (Summer): 52–71.

LaDuke, Winona. 2001. "Insider Essays: Our Responsibility." Electnet/Newswire, October 2. http://www.electnet.org/dsp_essay.cfm?intID=28. Last accessed Decemnber 22, 2002.

LaDuke, Winona, and Sean Aaron Cruz. 2013. *The Militarization of Indian Country.* East Lansing: Michigan State University Press.

Lee, Tanya H. 2014. "H-Bomb Guinea Pigs! Natives Suffering Decades after New Mexico Tests." Indian Country Today Media Network, March 5. http://indiancountrytodaymedianetwork.com/2014/03/05/guinea-pigs-indigenous-people-suffering-decades-after-new-mexico-h-bomb-testing-153856?page=0%2C3. Last accessed July 6, 2014.

MacMillan, Leslie. 2012. "Uranium Mines Dot Navajo Land, Neglected and Still Perilous." *New York Times*, March 31. http://www.nytimes.com/2012/04/01/us/uranium-mines-dot-navajo-land-neglected-and-still-perilous.html. Last accessed May 2, 2012.

Mead, Derek. "The U.S.'s Insane Attempt to Build a Harbor with a Two Megaton Nuclear Bomb; The AEC Almost Destroyed an Alaska Town in Its Attempt to Build a Nuclear Shovel. *Motherboard*, August 9, 2012. https://motherboard.vice.com/en_us/article/d777ak/the-u-s-s-insane-attempt-to-build-a-harbor-with-a-two-megaton-nuclear-bomb. Last accessed February 20, 2013. Last accessed March 31, 2013.

Nikiforuk, Andrew. 1998. "Echoes of the Atomic Age: Cancer Kills Fourteen Aboriginal Uranium Workers." *Calgary Herald*, March 14, A-1, A-4. http://www.ccnr.org/deline_deaths.html. Last accessed April 30, 1998.

"Prairie Island Indian Community Calls for Permanent Nuclear-Waste Solution". 2012. Indian Country Today Media Network, September 5. http://indiancountrytodaymedianetwork.com/article/prairie-island-indian-community-calls-for-permanent-nuclear-waste-solution-132820. Last accessed November 7, 2021.

Purdom, Tim. 2001. "Radiation Victims Rally; Marchers Complain about IOUs, Delays in Payments.' *Gallup Independent*, March 13. http://www.cia-g.com/~gallpind/3-13-01.html#anchor1 . Last accessed May 22, 2001.

Romero, Simon. "Why the Debate Over Russian Uranium Worries U.S. Tribal Nations." *New York Times,* May 2, 2022. https://www.nytimes.com/2022/05/02/us/us-uranium-supply-native-tribes.html. Last accessed May11, 2022.

Schneider, Keith. 1990. "Washington Nuclear Plant Poses Risk for Indians." *New York Times,* September 3, A-9.

Shaffer, David. 2012. "Prairie Island Tribe Seeks Action on Nuclear Waste." *Star-Tribune*, [Minneapolis] September 3. http://www.startribune.com/local/168337166.html?refer=y. Last accessed October 22, 2012.

References

"Uranium Miners, Families Bring Tales of Pain to Washington." 2000. *Arizona Republic*, April 15, n.p.

Vandegraft, Douglas L. Project Chariot: Nuclear Legacy of Cape Thompson." Proceedings of the U.S. Interagency Arctic Research Policy Committee Workshop on Arctic Contamination, Session A: Native People's Concerns about Arctic Contamination II: Ecological Impacts May 6, 1993, Anchorage, Alaska. http://arcticcircle.uconn.edu/VirtualClassroom/Chariot/vandegraft.html. Last accessed November 3, 1996.

Woody, R. L., B. Jack, and V. Bizahaloni. 1981. *Social Impact of the Uranium Industry on Two Navajo Communities*. Tsaile, AZ: Navajo Community College.

"Uranium/Nuclear Issues and Native Communities". 2001. Indigenous Environmental Network. http://www.ienearth.org/nuciss.html. Last accessed December 1, 2002.

Chapter 4
Showers of Pig Feces: A Neighborly Stench

Massive pig farms on an industrial scale produce lakes of swine waste that imperil the air and water of disproportionally Black, Native American, and other and poor people who endure frequent dousings of pig manure sprayed around their homes to fertilize nearby fields in, as Lily Kuo wrote in *Quartz* (2015), "a fine mist of liquefied feces [which] collects on their houses and cars, attracting swarms of flies." "The poor people, they literally get shit on," said Kemp Burdette, who advocates better water quality as an activist in North Carolina's Cape Fear River Watch (Kuo 2015).

In the language of agricultural bureaucracy, these gigantic pig farms are known as Swine CAFOs (Confined Animal Feeding Operations). The School of Public Health at the University of North Carolina—Chapel Hill reported that "these pig farms are responsible for both air and water pollution, mostly due to the vast manure lagoons they create to hold the enormous amount of waste from the thousands of pigs being raised for food" (Chiles 2015). North Carolina is one of several states in which the number of pigs exceeds the human population, and they produce several times more excrement per pig than human beings. No one seems to know exactly how much more (estimates range from two to 14 times).

4.1 Money Oinks at the State Capitol

Pig farming earns its proprietors billion dollars a year in North Carolina, and money oinks at the State Capitol, where legislators in April, 2017 passed a law limiting the damages that local residents might collect as a result of lawsuits alleging environmental racism—telling plaintiffs, in effect, that pigs and profits come before poor rural people, as Hog farms collect billions of gallons of untreated pig excrement in what are essentially cesspools, then dispose of the waste by spraying it into the air. Residents living in the sprayed areas complain of adverse health effects and a stench so wicked that it limits their ability to be outdoors (Goodman 2017).

Kuo (2015) described the creation of the waste lagoons: "The technology behind the lagoons is rudimentary at best. The hogs defecate in their indoor stalls, their waste falling through slatted floors into slurries where it is flushed or pumped into a nearby lagoon. Solid waste forms sludge at the bottom of the basin, creating a semi-permeable barrier that helps prevent leakage into the ground. Liquid from the top of the pond is used as manure and applied to crops with high-pressure spray guns. Unlike human waste, which is processed in municipal wastewater plants, the only treatment the pig manure receives is through exposure to naturally occurring bacteria in the lake."

Steve Wing, an associate professor of epidemiology at the University of North Carolina's Gillings School of Public Health, told Amy Goodman of the Public Broadcasting Service (PBS) that "The waste falls through the floors. It's flushed out into an open pit, like a cesspool. It's easy for a big hog operation to have as much waste as a medium-sized city. Of course, the pit will fill up, so it must be emptied.... by spraying the liquid waste.... And that can drift downwind into the neighboring communities" (Goodman 2017).

4.2 Wearing the Stench

Wing, who died during November 2016, described his research in a 2013 TED Talk: "In 1995, I began to meet neighbors of industrial hog operations," he said. "I saw how close some neighborhoods are to hog operations. People told me about contaminated wells, the stench from hog operations that woke them at night, and children who were mocked at school for smelling like hog waste. I studied the medical literature and learned about the allergens, gases, bacteria, and viruses released by these facilities—all of them capable of making people sick" (Hellerstein and Fine 2018).

The pig population in North Carolina rose more than 300% between 1990 and 2000, from 2.8 million to 9.3 million. Nationally, with about nine million pigs in 2017, North Carolina was second to only to Iowa among U.S. states. Duplin County alone housed 2.3 million hogs (Hellerstein and Fine 2018).

Shane Rogers collected evidence as an Environmental Protection Agency (EPA) and United States Department of Agriculture (USDA) environmental engineer and found that the manure spraying was literally bathing people's homes in pig-shit. He collected DNA swab samples from randomly selected houses, and "at every visit and every home, I experienced offensive and sustained swine manure odors [of] varying intensity, from moderate to very strong." Dust samples "contained tens of thousands to hundreds of thousands of hog feces DNA particles," Rogers wrote, "demonstrating exposure to hog feces bio-aerosols for clients who breathe in the air at their homes. Considering the facts, It is far more likely than not that hog feces also gets inside clients' homes where they live and where they eat" (Hellerstein and Fine 2018).

Elsie Herring, one local resident, said: "You'd think it's raining. We don't open the doors up or the windows, but the odor still comes in. It takes your breath away. Then you start gagging. You get headaches" (Goodman 2017). Don Webb, who shut down his own hog operation, told Amy Goodman of Democracy Now! on PBS: "I got out of it. And I couldn't—I just couldn't do another person that way, to make them smell [like] that. It is a cesspool that you put feces and urine in, a hole in the ground that you dump toxic waste in. And I've seen dead hogs in them and stuff like that. I've seen it. I've talked to the people. I've seen the little children that say, "mom and daddy, why do we got to smell this stuff?" You get stories like "I can't hang my clothes out. Feces and urine odor comes by and attaches itself to your clothes" (Goodman 2017).

The stench, and the way it has been ruining the lives of decent people turned Webb into an activist. "These are human beings," Webb said. "They've worked their whole lives and are trying to have a clean home and a decent place to live, and they can't go on their front porch and take a deep breath" (Hellerstein and Fine 2018).

The national and world news media beat a path to the hog farms of North Carolina to describe this putrid state of affairs; "'It smells like a decomposing body': North Carolina's polluting pig farms," Britain's *Guardian* headlined on its front page August 27, 2021.

4.3 A Stench that Smells like a Month-Old Decomposing Human Body

"It smells like a body that's been decomposed for a month," said Rene Miller, a retired truck driver from Duplin County. On Sundays after church, her family used to gather under an oak tree beside their house. They would dance, play checkers, and eat fried chicken, collard greens and corn. "That was my life back then," she said. Now, with hog waste sprayed onto a field across the road, she stays inside with the air-conditioning cranked up. (Yeoman 2019).

The stench combines the odors of methane, ammonia, rotten eggs, and spoiled collard greens. Some residents find themselves vomiting after breathing the rotten brew. Rene Miller, who lived nearby, described the stench as "an odor so noxious that it makes your eyes burn and your nose run. [as if it is] being surrounded by spoiled meat," a smell that grows worse on hot, humid summer afternoons, "when the stink hovering in the stagnant, humid air can nearly 'knock you off your feet'" (Hellerstein and Fine 2018). "I want to sit out in the front porch today but I can't because of the [pig-shit] spray," Miller said, adding how it's "disgusting" that she sometimes walks inside her home covered with a layer of fetid moisture from the spray (Strassmann 2016).

A 2018 study published in the *North Carolina Medical Journal,* as described in *The Guardian,* concluded that families living near huge hog farms suffered disproportionately high rates of infant mortality and deaths from anemia, kidney disease, and tuberculosis. Another inquiry, from 2014, found that these issues "disproportionately affect" people of color. African Americans, Native Americans, Latinos, and other people of mixed non-white ancestry are far more likely to live near these gigantic pig farms.

Such allegations brought forth loud denial from mangers of the pig farms. "We don't think these types of symptoms or things are going on in the communities where we do business," Kraig Westerbeek, a senior director at Smithfield Foods, the world's largest pork producer and the state's dominant player, said in a deposition in reference to reports that had found increased depression, anger, and confusion among neighbors who experienced hog-farm stenches. "There are studies that can say almost anything," he added (Yeoman 2019.)

4.4 So Much for Free Speech

Meanwhile, the state legislature passed some new laws meant, on the surface, to deal with the problems. However, existing farms were usually allowed to continue operating as usual. Environmental activists often were prohibited from making their complaints public, "to protect farmers from false accusations" (Yeoman 2019). So much for freedom of speech. The legislature was essentially telling people in pig-shit alley: smell the stench, be happy.

The entire charade continued as U.S. Environmental Protection Agency (EPA) investigators visited North Carolina and attempted to take statements from more than 60 hog-farm neighbors who described "strong stenches that made them gag, vomit, and lock themselves indoors." Several also mentioned keeping silent, because "for more than 15 years, the government has been well aware of the conditions they have to live with, but has done nothing to help, so complaining to [the] North Carolina DEQ [Department of Environmental Quality] would be futile." The EPA wrote in a letter to DEQ. Those who did complain reported "threats, intimidation, and harassment" by the industry (Yeoman 2019).

Twelve miles from Miller's home, Jeff Spedding, who works at Smithfield Foods, said that farmers want to be good neighbors, but he sees no better way to get rid of the waste. "I've never had a complaint from any of my neighbors. We try to do what is right," he said. "It [also] gets into what's cost-effective. It also gets into what's reasonable. There isn't any technology that's more efficient than what we're doing" (Strassmann 2016).

4.5 More Health Problems Related to Hog Waste

Remnants of hog cesspool waste have been rising into some residents' water wells. "It can, I think, very correctly be called environmental racism or environmental injustice that people of color, low-income people, bear the brunt of these practices,"

4.5 More Health Problems Related to Hog Waste

said Wing (Goodman 2017). "These lagoons and spray fields cannot be allowed to continue in North Carolina, they are causing too many problems to our waters, our air, our people, our health. They have got to go," said Rick Dove, a local activist with the Waterkeeper Alliance in New Bern, North Carolina and long-time campaigner against the lagoons (Kuo 2015).

The contamination from North Carolina pig farms has yielded dangerous levels of groundwater nitrates, a leading cause of births with heart defects (blue-baby syndrome). Occasionally, massive amounts of hog waste break out of containment ponds and flood nearby waterways. During 1995, six major hog-farm related spills totaling about 30 million gallons were mixed with water that is used for other purposes, killing more than 15 million fish.

The situation in North Carolina is not unique. Witness Uniontown, in northern Alabama, 90% Black and overwhelmingly poor, which provides a final resting place for the refuse of 33 states. "'Landfill' is too clean a word for what they do there," wrote Imani Perry in *Harper's*. "As part of Uniontown's sewage system, liquid waste is spewed into the air to land on the hard Alabama clay earth. The town is showered in shit" (2018, 61–62).

It's not just the pig-shit showers that bother locals in North Carolina. It's also the "dead boxes" filled with rotting hogs near a cemetery that attract gnats, swarms of large black flies, and buzzards. People loathed the ruination of their homes. Bumper stickers appeared, reading: "Welcome to North Carolina: Heaven 4 Hogs, Hell 4 Humans" (Hellerstein and Fine 2018). Among neighbors of these pork factories, hydrogen sulfide has provoked noticeable increases in respiratory ailments.

Steve Wing's work indicated that air pollution from nearby hog farms was related to higher-than-usual rates of nausea, high blood pressure, wheezing, and asthma, especially for children. Wing also coauthored a study that indicated contamination of nearby streams with fecal bacteria levels higher than state and federal water-quality standards allowed, according to regulations. Many of the same samples also showed excessive levels of *E. coli* and *Enterococcus*. Hog waste can also carry viruses, parasites, and even the "super-bug" MRSA.

C.D. Heaney and colleagues found that: "Testing of 187 samples showed high fecal indicator bacteria concentrations at both up- and downstream sites. Overall, 40%, 23%, and 61% of samples [from different sites] exceeded state and federal recreational water quality guidelines for fecal coliforms, E. coli, and enterococcus, respectively....Results suggest diffuse and overall poor sanitary quality of surface waters where swine CAFO density is high" (Heaney et al. 2015, 676).

Kate Jenkins wrote (2015): "Fishermen have reported that after exposure to the water, their skin will sometimes develop the same sores that tend to show up on dead fish." These are all compelling reasons to avoid swimming and fishing in contaminated streams. "When I used to go fishing, it used to be nothing to wade in that water... in tennis shoes, or barefoot! Would I do that now? No way!" [she was told by one resident] "If I catch a fish, you rest assured, it will be a catch and release."

Jenkins continued: (2015): "The life of each community member I meet seems in some way defined by serious illness. Stories of late-term miscarriage, brain tumors, unexplained childhood diseases, and death[s] punctuate[d] my two-day stay in hog country. These stories are not offered as evidence of anything — they are told offhand, in response to casual inquiries as to the speaker's well-being..."

4.6 Pig Manure, Race, Class, and Corporate Control

Wing and fellow UNC researcher Jill Johnston said that "The state's industrial hog operations disproportionately affect African Americans, Hispanics and Native Americans." That pattern, they concluded, "is generally recognized as environmental racism" (Hellerstein and Fine 2018). United States Senator Cory Booker, a New Jersey Democrat, denounced the North Carolina hog industry, which he called "evil" for exploiting its African American neighbors. "They fill massive lagoons with [waste] and they take that lagoon stuff and spray it over fields," he said: "I watched it mist off the property of these massive pig farms into Black communities. And these African American communities are like, 'We're prisoners in our own homes.' The biggest company down there [Smithfield] is...Chinese-owned, and so they've poisoned Black communities, land value is down, abhorrent ... This corporation is outsourcing its pain, its costs, onto poor Black people in North Carolina" (Hellerstein and Fine 2018).

Rather than stop the pig-shit showers, Republican State Rep. Jimmy Dixon, a farmer whose political campaigns received about $100,000 from the hog industry over five years, introduced the Agriculture and Forestry Nuisance Remedies bill, to block a class-action suit by nearly 500 mainly black residents seeking compensation from Murphy-Brown, North Carolina's largest pork producer. The class-action suit was filed by the North Carolina Environmental Justice Network, Waterkeeper Alliance, and REACH (Rural Empowerment Association for Community Help) under the Civil Rights Act of 1964, which prohibits governmental agencies from doing business in a way that has a disproportionate impact on low-income communities.

"These claims are at best enormous exaggerations and at worst outright lies. Is there some odor? Yes. But I would like you to close your eyes and imagine how ham and sausage and eggs and fried chicken smell," Dixon told a statehouse hearing on the bill, which was passed by both houses of the North Carolina legislature and forwarded to Gov. Roy Cooper for his signature. Cooper vetoed the bill, saying that he opposed "special protection for one industry" (Hellerstein and Fine 2018). The General Assembly then over-rode the veto during July 2017 by a vote of 74 to 40, mostly along party lines (Goodman 2017). Under HB 467, homeowners may be compensated only for *future reduction* in their property's fair-market value, which already has been substantially reduced by the presence of the hog farms. The law also negated existing legal proceedings and prohibits future lawsuits.

Corporate control of state politics is an important reason why North Carolina has become a magnet for the corporate pork industry. *The Guardian* reported that:

Nobody was more influential in reshaping the industry than Wendell Murphy, a powerful Democratic state legislator and the subject of the *Charlotte News & Observer's* Pulitzer Prize-winning "Boss Hog" series. Murphy, a high school agriculture teacher turned farmer from Rose Hill, grew to become the nation's top hog producer during his tenure in the general assembly, from 1982 through 1993. While in office, he backed legislation to provide poultry and hog farmers with tax breaks and exemptions from environmental regulation, helping "pass laws worth millions of dollars to his company and his industry," the *News & Observer* reported. This included the 1991 "Murphy Amendment," which exempted poultry and animal operations from stricter regulations on air and water pollution, and a 1991 bill that barred counties from imposing zoning restrictions on hog farms. In 1986, he voted in favor of a bill that eliminated sales taxes on hog and poultry operations. (Hellerstein and Fine 2018).

The hog industry, on the other hand, argues that few studies show a direct link between hog CAFOs and harm to people's health or the environment. Pollutants can also come from other sources, such as fertilizer, municipal human waste, or wildlife, they asserted. The industry maintains, in the face of copious contrary evidence, "that the lagoons, if maintained properly, can last forever and pose little threat to the overall environment or the health of nearby residents" (Kuo 2015).

In 2013, a Chinese firm, WH Group, acquired Smithfield Farms because the US' industrial model is 50% less expensive than Chinese hog production, which utilizes thousands of small farms. It is one area in which the United States easily under-sells the rest of the world. That worries many of their neighbors, who are tired of the stench and manure showers. Even Smithfield's corporate owner in China uses what lagoon skeptics would call more advanced technology that North Carolina lawmakers deemed too expensive to force farm compliance (Kuo 2015).

Problems provoked by daily life amidst the pig farms paled beside the destruction caused during September of 2018 by Hurricane Florence, the first Category 4 storm to hit North Carolina after nearly all of the huge-scale farms had been built. The remains of the hurricane stalled for several days over pig-farming regions, unleashing much as 40 inches of rain that flooded animal waste lagoons and industrial manufacturing plants on a scale never before recorded. The former pig-shit showers ceased for a while as the entire region became a gigantic hurricane-fed toilet as waste lagoons overflowed and mixed with floodwaters.

4.7 "When Pigs Fly"

In 2017, hopes were raised when the incumbent Republican governor, Pat McCrory, was outvoted by Democrat Roy Cooper, who appointed a former Environmental Defense Fund official Michael Regan to head DEQ. "We all had high hopes," said Naeema Muhammad, the Environmental Justice Network's organizing co-director said. The DEQ then issued "notices of deficiency" to nine large pig farms. Complaints now were being made public. The first posting covered November 2018 through April 2019, indicating that the DEQ had received 138 complaints and found 62 violations. The people living around the pork plants organized, filed several lawsuits against Smithfield, and won several hundred thousands of dollars in damages. The State Legislature, ever eager to please the mega pig farms' owners, then passed laws limiting its "neighbors'" ability to collect damages in the future.

Regardless of what has happened in the courtrooms, which Smithfield was planning to appeal, the "neighbors" of the pig farms as of 2022 were still facing the now world-famous stench. The publicity had not done anything to abate the stinking shit showers, which management of the gigantic pig farms continued relentlessly. The stench from the fields had not changed much from homeowners' front porches. "We have to deal with whether it's safe to go outside" said Elsie Herring, from her house within smelling distance of Smithfield's plants. It's a terrible thing to open the door and face that waste. It makes you want to throw up. It takes your breath away, it makes your eyes run," said Herring (Sainato and Skojec 2020).

Smithfield even kept a tally of the effluvia that its plants collected from its pools of pig feces, urine, and blood that gives the entire area its distinctive stench. Nine million pigs, a number close to North Carolina's 10.7 million people, supplies 15.5 million tons of feces per year. US citizens consume 18 pounds of bacon each on average. North Carolinians' bacon consumption is 47% above that—a lot of pig, but markedly below the national champion Nebraska, at 132% over the average. (These statistics were gathered June 25, 2022 from Google under the prompt: "Which state eats the most bacon?"). As of this writing, Omaha's member of the U.S. House of Representatives is named Don Bacon. I pig-shit you not.

Back at Smithfield in North Carolina, an effort has been underway to increase pigs' profit potential. The latest public-relations blitz as of 2022 was a prospective plan to turn Smithfield's fecal bonanza into methane gas. Thus, Smithfield hopes to capitalize on the pigs' ability to produce copious waste during their short lifetimes as a source of power for a methane-powered energy source. If it does nothing else, such talk gives the company cover to present the stinking mess as another avenue to profit as a power source. I would say "earth-friendly" power source, but methane is a greenhouse gas second only to carbon dioxide in potential to fry the planet via global warming.

However, in the eyes of the company, who are any no means atmospheric scientists, the stinking waste pools, with a whisk of a public-relations magic wand, become environmental assets. Given the lack of results thus far, the betting line on nearby front porches has been: "When pigs fly."

The whole idea disgusts and scares local residents and environmental activists, "who see it as seeking to profit from an ecological problem rather than fix it." Said Herring: "It only lines their pockets. They're trying to sell it as renewable energy. It's only renewable if pigs continue to poop, which is why I'm afraid they're going to [oppose a] moratorium on new hog farms, because if you have that great of a demand, you have to supply to meet it. They're not treating the waste, they're converting it, so how is that hog waste ever clean?" (Sainato and Skojec 2020).

Smithfield, as of this writing, had filed for a permit with the state Department of Environmental Quality for an industrial-scale biogas project, with 30 miles of pipelines to service the US's number-one source of commercial pork, which would cap waste lagoons, then conduct the methane through pipelines to a processing plant. The result, biogas, is part of a proposed $500 million joint venture between Smithfield Foods and Dominion Energy (Sainato and Skojec 2020). It is a stinking pity: all of this hot air and eco-hype to justify production of a greenhouse gas.

No sale, say many area residents, including Naemma Muhammad, a resident of Duplin county, who said: "Biogas is a false solution. It doesn't solve the problems we've been dealing with for three decades, which is to get rid of the lagoons and spraying systems so people can breathe and enjoy their property.... We don't need anything to encourage this industry to continue business as usual" (Sainato and Skojec 2020).

By 1990, North Carolina had the largest pig slaughter in the world. Activists had achieved some regulation of the pig farms' most noxious elements, battling political factions in the statehouse, only to learn that hog-farm interests have been whittling down the state's enforcement capabilities. One inspector, for example, was assigned many more farms to watch than was humanly possible.

4.8 Now Comes a Stampede of Chickens and Turkeys

At least as important in North Carolina agriculture was another type of farming that was growing much more rapidly than pigs. Noticing how profitable pigs had become with modest oversight, chicken and turkey producers have been moving in, processing more than 500 million poultry a year, many more than the nine million hogs (of course, hogs are much larger, so direct comparison is tricky). Chicken and turkey farms had grown much faster without even the slightest regulation, often in close proximity to pigs. While chickens and turkeys are less noxious than pigs per capita, 500 million of them pose a substantial waste potential of their own.

Melba Newsome wrote in Yale E360 (reprinted in *Mother Jones,* February 2022): Two eastern North Carolina counties, Duplin and Sampson, became synonymous with environmental racism when it came to light that the lower-income residents of they are largely Black, American Indian, and Latino counties were outnumbered by hogs 40 to 1. Even in North Carolina, in America's most densely populated swine-raising areas, "hog is no longer the boss of industrial farming," Newsome wrote (2022). "After massive expansion in recent years," he said, "poultry now trumps pig production in scale and economic impact and is increasingly seen as a threat to the environment and human health, chiefly because of the runoff of poultry waste into the state's waterways."

While activists advocated a moratorium on new hog farms, massive chicken and turkey mega-farms were thriving in many of the same areas. As long as people were buying them, and profits were to be made, animal protein was proliferating. Many people had quit eating pork for environmental reasons, but many of them continued to consume poultry. Soon, poultry pollution outnumbered that of pigs. By 2021, eastern North Carolina had twice as many poultry operations as pig producers.

Poultry manure added almost five times as much nitrogen and about four times as much phosphorus annually as pig manure in about 6500 CAFOs. Most of the chicken manure was dry, while pig faces are liquid, so its pollution is less obvious. Even so, chickens and turkeys were adding two million tons of waste to the local environment in an average year by 2021. Newsome (2022) described *en masse* as "a waste tsunami of bacteria-laden feces, urine and animal parts [that] poses a serious threat to North Carolinians' health, quality of life and property rights [a] state regulators have largely looked away" (Newsome 2022).

4.9 Hogs Now Run Second to Poultry

And what was more, poultry farms were increasing and expanding, at that time adding about 60 new farms per year. North Carolina's number of poultry farms doubled between 2016 and 2018; by 2020, about 1000 new such farms were being erected per year (Newsome 2022). Poultry installations soon outnumbered those of hogs three to one, with next to no state oversight (Newsome 2022). Chicken waste also was being used as a fertilizer, which eventually polluted the water table, lakes, and streams. Yale E360 (2022) reported that North Carolina poultry farms produced about five million tons of waste annually—about five times more waste (and an estimated five times more nitrogen and four times more phosphorus) than hog farms. At least some of the poultry and hog waste ends up in algal blooms that can eliminate or greatly reduce water-borne oxygen, with a toll on fish and other aquatic creatures, as well as rivers, and lakes with excess nutrients and other pollutants.

"In people," wrote Newsome, "The blooms also elevate toxins and bacterial levels that can leach into water, and play a role in development of gastrointestinal illnesses, skin reactions, and neurological problems."

The chicken and turkey farmers' worst nightmare involves a hurricane that comes ashore on the North Carolina barrier islands, then stalls over land and rains torrentially over the flat inland plains, an area with the state's highest density of CAFOs. Hurricane Matthew did this in 2016, and Hurricane Florence provided a sequel in 2018. Several million chickens drowned in each storm. Both storms also washed copious quantities of ammonia-laced chicken litter into state waterways. According to work done at the University of Arizona and the Nature Conservancy, 91 hog CAFOs and 36 poultry CAFOs sustained substantial damage from flooding. State environmental agencies are so hands-off with regard to poultry farms that a list of them is not available (Newsome 2022). There are, as a result, no inspections conducted nor permits issued. New poultry farms pop up like mushrooms after a spring rain, although start-up costs run about $1 million each.

It's enough to turn responsible people into vegans.

References

Chiles, Nick. "8 Horrifying Examples of Corporations Mistreating Black Communities with Environmental Racism." *Atlanta Black Star*, February 12, 2015. http://atlantablackstar.com/2015/02/12/8-horrifying-examples-of-corporations-mistreating-black-communities-with-environmental-racism/. Last accessed March 25, 2015. Last accessed January 10, 2016.

Goodman, Amy. "North Carolina Hog Farms Spray Manure Around Black Communities; Residents Fight Back." Democracy Now! May 3, 2017. https://www.democracynow.org/2017/5/3/nc_lawmakers_side_with_factory_farms. Last accessed June 2, 2017.

Heaney, C.D., K. Myers, S. Wing, D. Hall, D. Baron, and J.R. Stewart. "Source Tracking Swine Fecal Waste in Surface Water Proximal to Swine Concentrated Animal Feeding Operations." *Science of the Total Environment*. April 1, 2015. 676–683. https://www.ncbi.nlm.nih.gov/pubmed/25600418. Last accessed May 14, 2015.

Hellerstein, Erica and Ken Fine. "A Million Tons of Feces and an Unbearable Stench: Life Near Industrial Pig Farms." *The Guardian* (London, U.K.) September 20, 2017. https://www.theguardian.com/us-news/2017/sep/20/north-carolina-hog-industry-pig-farms. Last accessed May 11, 2018.

Jenkins, Kate. "Industrial Hog Farming and Environmental Racism." STIR. December 20, 2015. http://www.stirjournal.com/2015/12/20/industrial-hog-farming-and-environmental-racism/. Last accessed February 21, 2016.

Kuo, Lily. "The World Eats Cheap Bacon at the Expense of North Carolina's Rural Poor." *Quartz*, July 14, 2015. https://qz.com/433750/the-world-eats-cheap-bacon-at-the-expense-of-north-carolinas-rural-poor/. Last accessed September 2, 2015.

Newsome, Melba. "North Carolina Poultry Frenzy: 500 Million Birds and "Zero Transparency [Brought] Vast Lagoons of Hog Waste—Now Unregulated Mountains of Chicken Poop." *Mother Jones,* February, 2022. Reprinted from Yale E360. https://www.motherjones.com/environment/2022/02/north-carolina-poultry-farms-hog-waste-regulations-environmental-problems/. Last accessed April 22, 2022.

Perry, Imani. "As The South Goes, So Goes the Nation." *Harper's,* July 2018, 60–66.

Sainato, Michael and Chelsea Skojec. "The North Carolina Hog Industry's Answer to Pollution: a $500m [million] Pipeline Project." *The Guardian,* December 11, 2020, n.p.

Strassmann, Mark. "North Carolina Hog Farms Accused of Putrid Pollution." CBS News July 4, 2016. https://www.cbsnews.com/news/north-carolina-hog-farms-accused-of-putrid-pollution/. Last accessed September 30, 2016.

Yeoman, Barry. "'It smells like a decomposing body':" North Carolina's polluting pig farms." August 27, 2019. https://www.theguardian.com/environment/2019/aug/27/it-smells-like-a-decomposing-body-north-carolinas-polluting-pig-farms. Last accessed October 6, 2019.

Chapter 5
An Ice World Melts

Climate change induced by human consumption of fossil fuels impacts everyone on Earth and has become the signature environmental issue of our time. Native peoples of North America, with their close connection to the Earth and subsistence styles of life, are among the first to be most intensely affected by a rapidly changing climate. This is most evident in the Arctic, which is the most quickly warming area on Earth, where an Inuit world adapted to life on ice is quickly melting. Alaskan Native communities also face climate-induced change, including relocation of entire coastal villages. Elsewhere in North America, Native water resources and food sources already have been significantly damaged by a warming climate.

5.1 Sweating in Iqaluit

July 29, 2001, was a very warm day in Iqaluit, on southern Baffin Island, capital of the semi-sovereign Inuit nation of Nunavut in the Canadian Arctic. The bizarre weather was the talk of the town. The urgency of global warming was on everyone's lips. Even as U.S. President George W. Bush faulted global warming for lacking "sound science," the temperature hit 25°C. in a community that nudges the Arctic Circle, following a string of days that were nearly as warm. It was the warmest summer anyone in the area could remember. (The record high for Iqaluit is about 26.0 C., or 79.0 F.) Travelers joked about forgetting their shorts, sunscreen, and mosquito repellant—all now necessary equipment for a globally warmed Arctic summer.

In Iqaluit (pronounced "Eehalooeet"), a warm, desiccating northwesterly wind racing downhill off mountains raised whitecaps on nearby Frobisher Bay and rustled carpets of purple Saxifrage flowers on bay-side bluffs, as people emerged from their overheated houses (which have been built to absorb every scrap of passive solar energy), swabbing ice cubes wrapped in hand towels across their foreheads. The high temperature, at 78°F., was 30° above the July average of 48, comparable to a

110 to 115-degree day in New York City or Chicago. The wind raised eddies of dust on Iqaluit's gravel roads as residents swatted slow, corpulent mosquitoes.

5.2 Warming in the Arctic and Antarctic: How High Is Spectacular?

In a period of rising temperatures, highs in the upper 70s F. in Iqaluit, on southern Baffin Island (which I witnessed during the last week of July 2001), was so extreme for that area that it still stood 20 years later. It was in league with a worldwide warm spell during mid-March of 2022, when temperatures 50 to 70°F. above average smashed one-day records in both the Arctic and Antarctic.

Antarctica's Dome C-ii logged 14°F. Friday, March 18, 2022, where the average is minus 45°: "That's a temperature that you should see in January, not March (January is mid-summer there). That's dramatic. Both University of Wisconsin meteorologist Matthew Lazzara and ice scientist Walt Meier said what happened in Antarctica on March 18, 2022, is probably merely a random weather event and not a sign of climate change. But if it happens again or repeatedly, then it might be something to worry about and consistent evidence of global warming, they said.

The Arctic and Antarctic warm spells were first reported by The *Washington Post*, and picked up by Seth Borenstein of the Associated Press, one of the main-line media's most experienced climate-change reporters. The Antarctic continent as a whole at that time was about 8.6°F. warmer than its baseline temperature between 1979 and 2000, according to the University of Maine's Climate Reanalyzer, based on U.S. National Oceanic Atmospheric Administration weather models. That 8-degree heating over an already warmed-up average is unusual; think of it as if the entire United States was 8° hotter than normal, Meier said. At the same time, the Arctic as a whole was 6°F. warmer than the 1979–2000 averages. By comparison, wrote Borenstein, the world as a whole had warmed, on average (not based on any single day) only 1.1 C. degrees above the 1979–2000 average. Globally the 1979–2000 average is about half a degree C. warmer than its twentieth century average. What makes the Antarctic warming really weird is that the southern continent—except for its vulnerable peninsula (which is warming quickly and losing ice rapidly) has not yet been warming much, especially when compared to the rest of the globe, Meier said.

Professional weather watchers were quick to point out that these two polar warm spells didn't "prove" global warming. They preferred adjectives such as "weird," "stunning," "freakish," and so forth, perhaps to short-circuit climate skeptics who would undoubtedly accuse them of asserting an epic change of climatic epochs on one day's worth of evidence, even when that evidence came from several different places at both poles and averaged 50–70° above average. A comparison to the weather in Omaha, Nebraska (or New York City), where the average summer high is 88, would yield a reading of 138 F. (at 50°F. above average) or 158° (70° above average). No, it is not sustained climate change, not yet, but it would be one hell of a hot day. (Omaha's all-time record high is 114 F.)

For an indicator that includes more than one day, Antarctica did set a record for lowest summer sea ice (on a record extending back to 1979), shrinking to 741,000 square miles in late February, the snow and ice data center reported, according to Borenstein. What likely happened was that "a big atmospheric river" pumped in warm and moist air from the Pacific southward, Meier said, and Borenstein reported. Overall, the Arctic has been warming two to three times more quickly than the rest of the Earth, as relatively warm Atlantic air flows northward off the coast of Greenland. In some parts of the Arctic, thunderstorms were reported as salmon showed up in fishing nets for the first time. At what point does all of this jump from a one-day daily event from mere "weather" to more dangerous and enduring "climate?"

5.3 When Is it "Weather?" When Is It "Climate?"

Borenstein reported that "weather stations in Antarctica shattered records...as the region neared autumn." The two-mile high Concordia Station was at 10°F., which is about 70° warmer than average, while the even higher Vostok Station hit a shade above 0°F. beating its record by about 27°, according to a tweet from extreme weather record tracker Maximiliano Herrera. The coastal Terra Nova Base was far above freezing at 44.6°.

Borenstein continued: "The situation caught officials at the National Snow and Ice Data Center in Boulder, Colorado, by surprise because they were paying attention to the Arctic where it was 50° warmer than average and areas around the North Pole were nearing or at the melting point, which is really unusual for mid-March, said Meier." "They are opposite seasons. You don't see the north and the south (poles) both melting at the same time," Meier told The Associated Press. "It's definitely an unusual occurrence." It's pretty stunning," Meier added. "Wow. I have never seen anything like this in the Antarctic," said the University of Colorado ice scientist Ted Scambos, who had recently returned from an expedition to the continent. "Not a good sign when you see that sort of thing happen," said Lazzara.

5.4 Vast New Lakes Created

Meltwater from shrinking glaciers is creating vast lakes that could eventually pose an enormous flooding threat, said research released in 2020. "Unsurprisingly, we found [that] those lakes are growing. What was surprising was [by] how much," said Dan Shugar, a geographer at the University of Calgary (Climate Change Creating 2020).

The fact that glaciers around the world are shrinking due to climate change is well established. What has not been so well studied is where all that water is going. In a paper published during August 2020 in *Nature Climate Change*, Shugar and his colleagues provided the first global assessment of how much water is contained in the so-called glacial lakes and how quickly that volume is increasing. That assessment was not possible until a few years ago, when computers became powerful enough to work through a world's worth of data and 250,000 satellite images in a very short time.

Glacial lakes form when meltwater from glaciers is prevented from draining by the ice itself. They form on top, in front, beside, or even underneath a glacier. They are growing at a rapid pace everywhere glaciers are found. Shugar and his colleagues estimate that the amount of water those lakes hold has increased by almost 50% since 1990. The total volume has been calculated to be 158 cubic kilometers of water. That's a cube of icy water almost 5.5 kilometers long, wide, and high. Many glacial lakes are located in thinly inhabited locales such as Greenland. Others are in places like the Himalayas, where they sit alongside villages and communities.

Canadian glacial lakes also are swelling. Their volume across the country, including those in the High Arctic, has increased about 20% and they hold about 37 cubic kilometers of water, Shugar said. The lakes in British Columbia and the Yukon have increased even more quickly, almost doubling in volume over the last 30 years to 21 cubic kilometers (Climate Change Creating 2020).

This is alarming scientists because as permafrost thaws, carbon dioxide and methane previously locked up below ground are being released. These greenhouse gases can cause further warming, and further thawing of the permafrost, in a cycle known as a positive feedback. The higher temperatures also cause land ice in the Arctic to melt at a faster rate, leading to greater runoff into the ocean where it contributes to sea-level rise.

5.5 An Out-of-Season Swarm of Flies

Paula McLean-Sheppard, a Nunatsiavut Government employee, said she has been startled to see flies arriving earlier and earlier in the spring. "I was driving my truck last year and there was this big black cloud that suddenly appeared on the windshield. I had to use my wipers to clear it, I didn't know what it was," she said. "It was the *flies*. It used to be the end of June before you saw your first fly. Now they're coming in May" (Mercer 2018).

New species are arriving offshore, including cormorants, sharks, and sea turtles, coming as water warms, bringing new types of food. Seals, an important source of food and skin for waterproof clothes, are moving further north as local sea ice melts. During the winter of 2017 and 2018, local sea ice did not solidify until February, several months later than previously.

"Scientists say the impact of climate change on the Inuit psyche is significant, and only just beginning to be understood. Social workers worry it is leading to increased rates of drugs and alcohol abuse, in a place where the suicide rate far outstrips national rates. The land is not just a surface for them. It's family, it's kin, it's part of you. Every aspect of Inuit culture grows from the land," said Ashlee Cunsolo, director of the Labrador Institute in Happy Valley-Goose Bay. "When you're the first generation that can't do that any more, that can't follow your ancestors on to the land, think about how devastating that can be from a cultural standpoint. It's a disruption to cultural traditions that have endured for hundreds of years" (Mercer 2018). Cunsolo is one of the leading researchers into the links between climate change and mental health. She says the rapid changes happening in coastal Labrador are causing the Inuit to feel increased feelings of anxiety, depression, and grief. They sense something is being lost, she said. "These changes are disrupting hundreds of years of knowledge and wisdom and connection to the land. That's a scary thing for humanity," Cunsolo said (Mercer 2018).

5.6 Climate and Cultural Change

Is it possible to measure the degree to which culturally significant activities are changing due to climate change? Donna Hauser, an assistant professor at the International Arctic Research Center, University of Alaska at Fairbanks and colleagues work with people in Inuit communities, including several elderly advisers in Kotzebue, to describe changes in the hunting season as climate warms.

Their studies indicate that springtime hunts of bearded seals in Kotzebue ended around three weeks earlier in 2019 than in 2003. Sea-ice records indicate that that today, weather in late spring is similar to mid-summer in the mid-twentieth century. Klemetti Näkkäläjärvi, a postdoctoral researcher at the University of Oulu, has observed how cultural practices of his people, the Saami, has been threatened by climate change. Saami people live across Sápmi, from northern Norway into northwestern Russia.

It is possible to adapt reindeer herding as a livelihood or source of income to overcome the challenges posed by climate change. However, herding as a cultural form (that is, unchanged for a very long time—is vulnerable," Näkkäläjärvi asserted. "The greening process [less ice and snow] makes it difficult to navigate, travel, identify, search for and herd reindeer," explained Näkkäläjärvi (Baraniuk 2021).

Warmer waters also change the habitats of fish, which migrate to water temperatures that they find comfortable. Similarly, fish migrate northward. One good example is several species of salmon, which have been swimming northward through the Bering Strait into the Arctic Ocean. Näkkäläjärvi said that some Saami words have been disappearing because conditions to which they refer do not occur anymore. "No longer can a herder always trust his or her knowledge, the knowledge that has accumulated during centuries and that has safeguarded the success of Saami reindeer culture in harsh conditions to modern times," he added (Baraniuk 2021). The Nenets people, who herd reindeer in the Russian Arctic, have also experienced such difficulties. They are used to migrating for hundreds of miles with their animals to find pastures to graze but climate change, and the associated rise of oil and gas infrastructure in this part of the world, has made this incredible journey much more difficult.

"Because temperatures in the Arctic are rising faster than anywhere else in the world, we must look to the experiences of Inuit as a harbinger of what is to come—and seek their guidance on how to live sustainably. Indigenous peoples of our country, have already experienced life-threatening emergencies on many levels, and are now at the front lines of the slow, multifaceted disaster that is climate change. Virtually every community across the North is now struggling to cope with extreme coastal erosion, thawing permafrost, and rapid destructive runoff, which particularly affects coastal communities in Alaska and in northern and western Canada." wrote Sheila Watt-Cloutier, in *Canadian Geographic* (November 15, 2018). Watt-Cloutier was a 2007 Nobel Peace Prize nominee, and author of *The Right to Be Cold,* (2015). She is Inuk and one of the most widely respected political figures to emerge in the Arctic.

5.7 Sea Ice Is in Rapid Decline

Watt-Cloutier continued: "Despite our cold northern winters, sea ice remains in rapid decline. Glacial melt, long relied on for drinking water, is now unpredictable. In one stunning case, the Kaskawulsh Glacier in the Yukon has receded so far that its meltwater has changed direction, flowing south toward the Gulf of Alaska and the Pacific Ocean instead of north toward the Bering Sea. Ice that used to serve as our winter highways is giving way and invasive species are travelling much further north than ever before. While the impact and extent of each change varies across the North, the trends are consistent. The change is not just coming; it is already here" (Cloutier, *Canadian Geographic*, November 15, 2018).

Arctic ice is the Earth's air conditioner. It is melting, and that air conditioner is breaking down. Inuit have much wisdom to share with the world about living sustainably, in harmony with nature—all while coping with the effects of climate change. Cloutier concluded: (November 15, 2018): "The whole planet benefits from a frozen Arctic and Inuit still have much to teach the world about the vital importance of Arctic ice, not only to our culture, but to the health of the rest of the planet."

Inuit people see their climate not as extreme but as demanding (Sansoulet et al. 2020; Therrien 2007). They have learned to adapt over the centuries; their culture has evolved to survive in the Arctic. Too much climate change too quickly (for example, too much warming that melts ice too quickly) can damage the economy of the Inuit, especially if temperatures rise to levels that people from lower latitudes find comfortable.

Today's Inuit report that weather in their homelands has become less predictable as well as warmer than in the previous 30 years (Sansoulet et al. 2020; Weatherhead et al. 2010). Impacts of these changes may significantly affect Inuit health, harvesting activities, and ecosystems.

The ice is changing; the flatness and the bumps are different than when I used to hunt. The ice is melting faster than it used to be, back then in [the] 1970s. (...) Floe ice does not seem to come early anymore, because of the wind. (...) Faster melting and slower freezing (Inuit elder from Kanngiqtugaapik).

Inuit from several communities reported changes on the ice as one of the major changes which have occurred (Video S1). The stability of ice is changing, as it freezes later in the fall than previously.

Back then, if we were in [the] Arctic if it was a winter it would be minus 60 F. but...these days it seems to be going [to] minus 40 F. and that environment seems to be changing [...] It's definitely getting warmer. And when someone is trying to build one [an igloo] they're having a hard time finding the right type [of snow]. You know texture of the snow is different now. The snow is softer now (Inuit adult from Kanngiqtugaapik).

These changes in stability of timing of melting pose dangerous problems for hunters. They have been forced to alter usual routes to new cracks in the ice. Such route changes can lengthen distances traveled by hunters to reach the best areas for hunting. When icescapes change in unforeseen ways, uncertainty also rises, along with a general feeling of danger rooted in uncertainty about hunters' safety, spurring impacts on their psychological state of hunters and their families (Dowsley et al. 2010). Successful, safe, and secure hunting depends on knowledge of patterns in ice flow and stability. Hunters keep mental records of where ice cracks from, strength and direction of winds, and other factors that allow them to gauge the safety of the sea ice with regard to hunting and transportation. Rapid changes in climate affect the entire frame of Inuit reality. Adaptation to unpredictable and sometimes previously unknown conditions makes hunters' lives increasingly more difficult.

Such changes have brought a decline in the presence of caribou in Pangniqtuuq (which means "the place of many caribou bulls") to seek other species to eat, such as ptarmigan.

Everybody tries to catch it [ptarmigan] and we never stop to catch it, everybody wants it [be]cause it's good cause we're not supposed to catch caribou and we make soup out of the ptarmigan. It's almost the same taste of caribou, the rabbits too, almost same taste of caribou meat. Because they're eating the same thing. (Inuit elder from Pangniqtuuq).

When he says, "we're not supposed to catch caribou," he is referring to the implementation of quotas by the government, which regulate the number of caribous each community may hunt in a given year. Both of these quotas and changes in the caribou migration lead to a significant decrease in caribou consumption.

5.8 Food Insecurity and Grocery Stores

Inuit hunters and families usually have a clear preference for natural, traditional food, but the so-called store-bought food has become popular, especially when little else is available. At that time, Inuit people find themselves unwillingly sucked into the money economy, given that imported food is uncommonly expensive because it must be brought in by air. Spoilage rates are high, and people become dependent on high sugar, high fat, sugary, processed food. Resulting food insecurity is particularly high in Nunavut, where 68.8% of adults experience a high level of it, compared to a national Canadian average of 9.2% (Huet et al. 2012).

As an Inuit elder from Kanngiqtugaapik explained, food insecurity is attributable to the arrival of supermarkets, such as Northern (which is owned by the North West Company), because "It is easier," requiring less physical effort to acquire food.

If I'm working, I'll buy more stuff at Northern [a food store]. *If I'm not working, I get more from outside. Because sometimes I work seven days a week, 11 hours a day. But spring and summer, 24-hour daylight, after working hours, I'm gone. I only got few hours to rest, come back, cut my seal, go to bed and in the morning, go to work. (Inuit elder from Kanngiqtugaapik).*

In the Eskimo village of Kaktovik, Alaska, on the Arctic Ocean roughly 250 miles north of the Arctic Circle, a robin built a nest in town during 2003—not an unusual event in more temperate latitudes, but quite a departure from the usual in a place where, in the Inupiat Eskimo language, no name existed for robins, until recently. Some residents of Baker Lake, Nunavut, 1330 kilometers west of Iqaluit, spotted magpies flitting around town during May 2006. These scavengers, a relative of the crow, had never been seen in Nunavut before. The Magpies are not expected to become permanent residents, however, even if the climate warms, because they roost in trees. The tundra has no trees—until changes in climate provides them. In the Okpilak River Valley, which heretofore has been too cold and dry for willows, they have sprouted profusely. Never mind the fact that and in the Inupiat language *Okpilak* means "river with no willows." Three kinds of salmon have been caught in nearby waters in places where they were long unknown, within living memory.

5.9 "We Have Never Seen Anything Like This. It's Scary, *Very* Scary"

Correspondent Jerry Bowen, airing on the C.B.S. Morning News August 29, 2002, from Barrow, Alaska, the northernmost town in Alaska, quoted Simeon Patkotak, a native elder, saying that residents there had just witnessed their first mosquitoes. Ice cellars carved out of permafrost were melting as well, forcing local Native people to borrow space in electric freezers for the first time to store whale meat. Average temperatures in Barrow have risen 4°F. during the past 30 years (Bowen 2002). The average date at which the last snow melts at Barrow in the spring or summer has receded about 40 days between 1940 and the years after 2000, from early July to, some years, as early as mid-or-late May (Wohlforth 2004, 27).

"We have never seen anything like this. It's scary, *very* scary" said Ben Kovic, Nunavut's chief wildlife manager. "It's not every summer that we run around in our T-shirts for weeks at a time". At 11:30 a.m. on a Saturday in late July, Kovic was sitting in his back yard, repairing his fishing boat, wearing a T-shirt and blue jeans in a mild wind, with many hours of Baffin's eighteen-hour July daylight remaining. On a nearby beach, Inuit children were building sandcastles with plastic shovels and buckets, occasionally dipping their toes in the still-frigid seawater.

Addressing a U.S. Senate Commerce Committee hearing on global warming on August 15, 2004, Watt-Cloutier, then president of the Inuit Circumpolar Conference, based in Inuit, said: "The Earth is literally melting. If we can reverse the emissions of greenhouse gases in time to save the Arctic, then we can spare untold suffering." She continued: "Protect the Arctic and you will save the planet. Use us as your early-warning system. Use the Inuit story as a vehicle to reconnect us all so that we can understand the people and the planet are one" (Pegg 2004).

The Inuits' ancient connection to their hunting culture may disappear within her grandson's lifetime, Watt-Cloutier said. "My Arctic homeland is now the health barometer for the planet" (Pegg 2004). Committee chair John McCain, an Arizona Republican (and later a Republican candidate for U.S. president), said a recent trip to the Arctic showed him that "these impacts are real and consistent with earlier scientific projects that the Arctic region would experience the impacts of climate change at a faster rate than the rest of the world. We are the first generation to influence the climate and the last generation to escape the consequences" McCain said (Pegg 2004).

Warming has extended to all seasons. In Iqaluit, for example, thunder and lightning used to be an extreme rarity. The world of the Arctic is changing rapidly before indigenous peoples' eyes. Lightning strikes, for example, ignite fires on the tundra just as in forests; more than 20,000 were recorded on Alaska's North Slope in 2007, a period of record high temperatures and widespread drought in the area. One strike torched a fire near the Anaktuvuk River. By the time its embers were snuffed by falling snow in early October, it had scorched an area the size of New York City (Qiu 2009, 34). According to the U.S. Bureau of Land Management, the frequency of lightning on the North Slope has increased by a factor of ten in ten years (Qiu 2009, 34). Warming brings more lightning, and lightning sparkled fires add carbon dioxide to the air; tundra stores 14% of the world's surface soil carbon (Qiu 2009, 34). Fires also accelerate the thawing of permafrost, adding even more carbon dioxide and methane to the atmosphere.

The same day that the high hit 78°F. in Iqaluit, the forecast for Yellowknife, in the Northwest Territories, called for a high of 85°F. with scattered thunderstorms—a usual summertime forecast in Chicago or New York City. During the summer of 2005 and afterwards, highs in the 80s F. became routine in Fairbanks, Alaska. The winters of 2000–2001 and 2001–2002 in Iqaluit were notable for liquid precipitation (freezing rain) in December. A few years later, thunderstorms bearing heavy rain were reported during brief temperature spikes even in February across Alaska's North Slope, over and near Iqaluit, and along the southwestern coast of Greenland.

The harbor ice at Iqaluit did not form in 2000 and 2001 until late November or December, five or six weeks later than previously usual. The decline of the ice continued after that. In later years, the late onset of ice ceased to be unusual. The ice also breaks up earlier in the spring—sometimes even in late April or May in places that once were icebound into early July. Early in January 2004, Watt-Cloutier wrote from Iqaluit, on Baffin Island, that Frobisher Bay had just frozen over for the season at a record late date:

> We are finally into very "brrrrrr" seasonal weather and the Bay is finally freezing straight across. At Christmas time the Bay was still open and as a result of the floe edge being so close we had a family of Polar bears come to visit the town a couple of times. Also in Pangnirtung [north of Iqaluit] families from the outpost camps came into town for Christmas by boat! Imagine that the ice was not frozen at all there by Christmas. But this week our temperatures with the wind-chill reached minus 52, so we are pleased... (Watt-Cloutier 2004).

Watt-Cloutier looked out at the waters of Frobisher Bay two weeks after she had represented the Inuit Circumpolar Conference at a Conference of Parties to the United Nations Framework Convention on Climate Change in Milan, Italy, where she said, in part:

> Talk to hunters across the North and they will tell you the same story—the weather is increasingly unpredictable. The look and feel of the land is different. The sea-ice is changing. Hunters are having difficulty navigating and traveling safely. We have even lost experienced hunters through the ice in areas that, traditionally, were safe! Our Premier, Paul Okalik, lost his nephew when he was swept away by a torrent that used to be a small stream. The melting of our glaciers in summer is now such that it is dangerous for us to get to many of our traditional hunting and harvesting places....Inuit hunters and elders have for years reported changes to the environment that are now supported by American, British and European computer models that conclude climate change is amplified in high latitudes (Watt-Cloutier 2003).

Rapid warming has extended into mid-winter. During the first week of January 2011, record warmth in Iqaluit drove temperatures above the freezing point, an extreme rarity there in mid-winter, creating a debilitating, muddy mess. After that, everything flash-froze into dangerous sheets of ice. The temperature at one point was 68°F. above average. In New York City, a similar departure from average would have produced a 100 F. on New Year's Eve. A high of 35 F. in the first week of January was an all-time record. Nearby, the village of Pangnirtung hit 8 C (46 F) on January 4. "The normal around this time of year is around minus 22 C (minus 8 F)," Yvonne Bilan-Wallace, a meteorologist with Environment Canada, told the *Nunatsiaq News* in Iqaluit. "So yeah, you're way above normal" (Windeyer 2011). "Accustomed to blizzards, Iqaluit road crews are finding it challenging to keep the streets passable," reported the Indian Country Today Media Network. "When rain covered the previous day's sand, then froze, they couldn't put down more sand until they had chipped the ice away with a grader" (Nunavut Capital 2011). In the Inuit village of Coral Harbour, in the Canadian Arctic near the Hudson Bay's mouth, temperatures also exceeded the freezing point in January 2011 for the first time on record during that month. Hudson Bay could be ice-free by the middle of this century.

Alaska Natives (such as Yupik in St. Mary's and Pitka's Point) have told scientists that weather has become more unstable in ways that affect their lives. For example, less melting snow means that thawing rivers carry less of the driftwood that Native peoples use for cooking and heat. In Point Hope, Alaska, 330 miles south of Point Barrow, a 4500-year-old Alaskan native village of 900 people, wrote Juliet Eilperin (2012) in the Washington *Post*,

> Fermented whale's tail doesn't taste the same when the ice cellars flood.... Whaling crews in this Arctic coast village store six feet of tail—skin, blubber and bone—underground from spring until fall. The tail freezes slowly while fermenting and taking on the flavor of the earth. Paying homage to their connection with the frozen sea, villagers eat the delicacy to celebrate the moment when the Arctic's ice touches shore. But climate change, with its more intense storms, melting permafrost and soil erosion, is causing the ice cellars to disintegrate. Many have washed out to sea in recent decades. The remaining ones regularly flood in the spring, which can spoil the meat and blubber, and release scents that attract polar bears.

Stored in an electric freezer, the fermented whale's tail loses its distinctive taste. "So much of our culture is being washed away in the ocean," said Point Hope Mayor Steve Oomittuk. "We live this cycle of life, which we know because it's been passed from generation to generation. We see that cycle breaking" (Eilperin 2012). Melting ice deprives hunters of firm footing, and obliterates paths that used to be reliable. Instead of poking their heads through holes in the ice, seals remain submerged in water, evading hunters.

The cascading effects of climate change on North American Native peoples became the sole focus of an entire issue of the scientific Journal *Climatic Change* published in October 2013. Special attention was paid to the congruence between native elders' observations and meteorological records, finding widespread agreement. Native subsistence was a major focus of this scientific focus, from observations that "hunters have observed that moose seem less healthy" (Lynn et al. 2013, 549), to effects of diminishing sea ice on walrus, effects on salmon runs, and damage from increased storm intensity and rising seas. Some shore-dwelling tribes along the Washington coast (Quileute and Hoh, for example) have moved, or are considering leaving sea-front homelands inundated by rising storm surges.

Many Native residents of the Arctic have asserted that the weather with which they live has become stormier as weather has warmed. Late in 2009, two scientists in the School of Earth Sciences at the University of Melbourne, Victoria, Australia, provided statistics that support those assertions. Ian Simmonds and Kevin Keay studied the month of September because sea ice has been markedly reduced from 1979 through 2008, and found that cyclonic storms had become stronger, although the number of storms had not changed significantly. "The findings reinforce suggestions that the decline in the extent and thickness of Arctic ice has started to render it particularly vulnerable to future anomalous cyclonic activity and atmospheric forcing," they wrote in *Geophysical Research Letters* (Simmonds and Keay 2009).

Ice is fundamental to the entire Arctic food chain.

> Sea ice is the foundation of the Arctic marine environment. Vital organisms live underneath and within the ice itself, which is not solid, but pierced with channels and tunnels large, small, and smaller. Trillions of diatoms, zooplankton, and crustaceans pepper the ice column. In spring, sunlight penetrates the ice, triggering algal blooms. The algae sink to the bottom, and in shallow continental shelf areas they sustain a food web that include clams, sea stars, arctic cod, seals, walruses,—and polar bears (McGrath 2011, 70).
>
> The swiftly changing Inuit world revolves around the ice, which is melting rapidly year by year. The traditional economy and its spiritual bonds are tethered to the ice. The pronounced thinning of Arctic sea ice has made the ice pack more brittle and susceptible to wind drift. The volume of Arctic sea ice decreased by one-third during 2007–11compared with the 1979–2006 mean. As Richard Kerr wrote in *Science* (2012, 1591):
>
> The now-clearly-accelerating decline of summer ice—punctuated by exceptional losses in 2007 and now in 2012 [and through 2021 by later observers]—has persuaded everyone that summer Arctic sea ice will be a thing of the past far sooner than the end of the twenty-first century, as current models predict. So the full knock-on effects of an ice-free Arctic Ocean—from the loss of polar bear habitat to possible increases of weather extremes at mid-latitudes—could be here in many people's lifetimes. How far wrong the models might be, however, is still very much in dispute.

5.10 Worst Fears Realized

"It's hard even for people like me to believe, to see that climate change is actually doing what our worst fears dictated," said Jennifer A. Francis, a Rutgers University scientist who studies the effect of sea ice on weather patterns. "It's starting to give me chills, to tell you the truth" (Gillis 2012). Summer 2012's low 3.41 million square miles of water-laced ice was 18% below the former 2007 record (by an area as large as Texas), and half of the area measured by the first satellite photos in 1979. Between 1979 and 2001, ice declined in the Arctic on an average of 6.5% a year. In the next 11 years, the average rate of decline doubled. "The rules are changing," said Mark Serreze of the National Snow and Ice Data Center. "The ice is so thin, it just can't survive the summer" (Kerr 2012, 1591). In the nine years since Kerr said these words, the size of the Arctic ice cap in summer has continued to a point where the once-fabled Northwest Passage probably will open within a generation. The Arctic Ocean probably will lose its summer ice by the end of the twenty-first century.

"When the ice thins to a vulnerable state, the bottom will drop out and we may quickly move into a new, seasonally ice-free state of the Arctic," Serreze said. "I think there is some evidence that we may have reached that tipping point, and the impacts will not be confined to the Arctic region" (Arctic Sea Ice Melt 2008). The review paper by Serreze and Julienne Stroeve of CU-Boulder's NSIDC and Dr. Marika Holland of the National Center for Atmospheric Research titled "Perspectives on the Arctic's Shrinking Sea Ice Cover" appeared in the March 16, 2007 issue of *Science*. "We're seeing more melting of multi-year ice in the summer," said Stroeve. "We may soon reach a threshold beyond which the sea ice can no longer recover." "We have already witnessed major losses in sea ice, but our research suggests that the decrease over the next few decades could be far more dramatic than anything that has happened so far," said Holland. "These changes are surprisingly rapid" (Arctic Sea Ice Melt 2008). In August and September 2012, sea ice covered less of the Arctic Ocean than at any other time since at least 1979, when the first reliable satellite measurements began. In mid-September, Arctic sea ice had declined to 3.41 million square kilometers (1.32 million square miles), much less than the previous record of 4.17 million square kilometers (1.61 million square miles) in 2007. Melting sea ice feeds itself. With less ice to reflect sunlight, the sea dark surface absorbs more energy. Less ice survives year to year; new ice that freezes in winter is thinner and melts faster as the sun circles the Arctic sky in summer.

"It's an example of how uncertainty is not our friend when it comes to climate-change risk," said Michael E. Mann, a climate scientist at Pennsylvania State University. "In this case, the models were almost certainly too conservative in the changes they were projecting" (Gillis 2012).

5.11 Land of Melting Ice and Burning Tundra

As tundra and permafrost thaws, it releases stored carbon dioxide and methane which further raises the atmosphere's level of greenhouse gases. During unusually warm, dry winters, the snowless tundra often catches fire, a precursor of larger fires that could further accelerate global warming by adding carbon dioxide to the atmosphere. There exists a palatable fear among climate scientists that such "biotic feedbacks" in the oceans and on lands could release large amounts of carbon dioxide and methane which are now stored in the earth. All of this increases the speed with which the Arctic ice world is declining, and the lives of Arctic peoples are changing. In subsequent years, summer wildfires have become the normal state of affairs nearly everywhere that tundra intersects with the Artic Circle.

5.12 Warming's Pervasive Effects

A warming climate is reshaping daily life in the Arctic, including decisions about such basics as which streams or rivers are safe sources of drinking water. The beaver population has risen as the weather warms. "The rodent's desecration is suspected of heightening the risk of *giardia,* an intestinal infection often referred to as 'beaver fever,'" reported *Mother Jones* (Lee 2011). "In general, people could drink from [the creeks and rivers] freely. Now they have beavers defecating into the river," said Michael Brubaker, director of the health consortium's Center for Climate and Health (Climate Change Puts 2011).

Rising temperatures enhance the probability of disease in many fish on which Native peoples depend. For example, "Chinook salmon are now commonly infected with a single-celled parasite previously unknown in Alaskan salmon" (Lynn et al. 2013, 550). Salmon, a cold-water fish, is very sensitive to warming water, which debilitates and often makes them ill. In summer, 2003, "salmon paused their upstream migration, remaining below the Bonneville Dam on the Columbia River, for four weeks until temperatures cooled," noted one account (Cozzetto et al. 2013, 576). Warmer water is causing mussels to disappear in parts of Pacific Northwest waters. Rising ocean acidity provoked by rising levels of carbon dioxide in air and water also is imperiling shellfish. Along some parts of the Pacific Coast, shellfish have stopped reproducing because sea water contains too much carbon dioxide and has become unnaturally acidic.

In the Arctic, wrote Martin P. Tingley, and Peter Huybers in *Nature*:

> The summers of 2005, 2007, 2010 and 2011 were warmer than those of all prior years back to 1400, in terms of the spatial average. The summer of 2010 was the warmest in the previous 600 years in western Russia, and probably the warmest in western Greenland and the Canadian Arctic as well. These and other recent extremes greatly exceed those expected from a stationary [stable] climate (2013, 201).

Nick Jans, writing in *The Last Polar Bear*, sketched the ground-level effects of climate change in Barrow:

> A sharp, insistent wind sifts a rolling haze of snow along the ground, but for mid-December the temperature is positively balmy: plus 19°F. A generation ago, 20 or even 30 below zero would have been the norm. The last few days amount to more than some freakish warm spell; this year, the sea ice has yet to become shorefast. Not so many years ago, solid freeze-up occurred by the end of October, without fail. Now it's six, seven, even ten weeks later, and the ice proves thinner and less stable. The elders shake their heads, bemused as if palm trees had sprouted on the shores of the Beaufort Sea. They're beyond debating climate change. It's staring them right in the face (Jans 2008, 150, 153).

5.13 A Winter Without Walrus

Hunting, including walrus, is still key to survival for Native people across the Arctic. Imported food is not only culturally out of character, but also so expensive that few Native people can afford it. Wild swings in weather and ice conditions has devastated harvests of walrus, the main subsistence staple in some Alaskan Native villages. In recent years, ice, the walrus' primary habitat, has been too scarce—except, as in the winter of 2012–2013, when weather was so cold and ice so thick that hunters could not get out of their home villages often enough to provide for their families.

Food imported by air to the one small grocery store in Gambell, Alaska (in extreme northwest Alaska within sight of Siberia's Chukchi Peninsula) is no alternative. A single small chicken cost $25 in 2001. Families who once put 20 to 25 walruses in their meat lockers each spring to last through the winter have been storing only a handful in recent years because of retreating ice. "if this continues, we will seriously starve," said Jennifer Campbell, 38 years of age in 2013, mother of five, whose family caught only two walruses in the autumn of 2013, down from about 20 previously (Carlton 2013, A-4). The Yupik Eskimos harvest the walrus from passing ice floes in May and June using small boats. In 2013, however, the hunt lasted only a week. "The ice was so bad we couldn't get out," said Brian Aningayou, a hunter who took only four walrus to feed an extended family of 40 people (Carlton 2013, A-4). He has been shooting birds, which provide little meat compared to a walrus that can be the size and weight of a small car.

With ice receding hundreds of miles offshore of Alaska and Russia during the late summer of 2007, walruses gathered by the thousands onshore of Alaska and Siberia. Joel Garlich-Miller, a walrus expert with the U.S. Fish and Wildlife Service, said that walruses began to gather onshore late in July, a month earlier than usual. A month later, their numbers had reached record levels from Barrow to Cape Lisburne, about 300 miles southwest, on the Chukchi Sea.

Walruses usually feed on clams, snails, and other bottom-dwelling creatures, diving to a shallow bottom from the ice. In recent years, the ice has receded too far from shore to allow the usual feeding pattern. A walrus can dive 600 feet, but the depth of water under ice shelves in late summer is now several thousand feet deep. The walruses have been forced to swim much farther to find food. Finding less, they return to starving calves. More calves also are being orphaned. Russian research observers also reported many more walrus than usual on shore, tens of thousands in some areas along the Siberian coast, which would have stayed on the sea ice in earlier times (Joling 2007a, October 4). Walrus are prone to stampedes once they are gathered in large groups. Thousands of Pacific walruses above the Arctic Circle were killed on the Russian side of the Bering Strait, where more than 40,000 had hauled out on land at Point Schmidt, as ice retreated further northward. A polar bear, a human hunter, or noise from an airplane flying close and low can send thousands of panicked walrus rushing into the water.

"It was a pretty sobering year, tough on walruses," said Joel Garlach-Miller, a walrus expert for the U.S. Fish and Wildlife Service. Several thousand walrus died late in the summer of 2007 from internal injuries suffered in stampedes. The youngest and the weakest animals, many of them calves born the previous spring, were crushed. Biologist Anatoly Kochnev of Russia's Pacific Institute of Fisheries and Oceanography estimated 3000–4000 walruses out of population of perhaps 200,000 died, or two or three times the usual number on shoreline haulouts (Joling 2007b, December 14). "We were surprised that this was happening so soon, and we were surprised at the magnitude of the report," said wildlife biologist Tony Fischbach of the U.S. Geological Survey (Joling 2007b, December 14). Walrus lacking summer sea ice that they use to dive for clams and snails, may strip coastal areas of food, and then starve in large numbers.

5.14 An Oral History of a Melting World

Around the Arctic, in Inuit villages now connected by the oral history of traveling hunters as well as e-mail and the Internet, weather watchers are reporting striking evidence that global warming is an unmistakable reality. Weather reports from the Arctic sometimes read like the projections of the Intergovernmental Panel on Climate Change (I.P.C.C.) on fast forward. These personal stories support I.P.C.C. expectations that climate change will be felt most dramatically in the Arctic.

In the 1990s, Alaskan Native weather watchers established websites to share their experiences with climate change: "Turtles appearing for the first time on Kodiak Island, birds starving on St. Lawrence Island, thunder first heard on Little Diomede Island...snowmobiles falling through the ice in Nenana....Already, the central Arctic is warming 10 times as fast as the rest of the planet, outpacing even our attempts to describe it" (Frey 2002, 26). During the summer of 2004, several *Vespula intermedia* (yellow-jacket wasps) were sighted in Arctic Bay, a community of 700 people on the northern tip of Baffin Island, at more than 73° North latitude.

Noire Ikalukjuaq, the mayor of Arctic Bay, photographed one of the wasps at the end of August. Ikalukjuaq, who said he knew no word in Inuktitut (the Inuit language) for the insect, reported that other people in the community also had seen wasps at about the same time (Rare Sighting 2004).

Growing numbers of Inuit are suffering allergies from white-pine pollen that recently reached Kuptana's home in Sach's Harbour, on Banks Island, for the first time. "If this rate of change continues, our lifestyle may forever change, because our communities are sinking with melting permafrost and our food sources are ... more difficult to hunt". Sach's Harbour itself is slowly sinking during the summer into a muddy mass of thawing permafrost. Born in an igloo, Kuptana has been her family's weather watcher for much of her life (she was 68 years of age in 2022). Her job was to scan the morning clouds and test the wind's direction to help the hunters decide whether to go out, and what they should wear. Now she gathers observations for international weather-monitoring organizations.

5.15 The Inuit World Turned Upside down

By the winter of 2002–2003, warming weather sometimes was forcing hockey players in Canada's far north to seek rinks with artificial ice. Canada's *Financial Post* reported: "Officials in the Arctic say global warming has cut hockey season in half in the past two decades and may hinder the future of development of northern hockey stars such as Jordin Tootoo" (Ice is Scarce 2003). According to the *Financial Post* report, hockey rinks in northern communities were raising funds directed toward installation of cooling plants to create artificial ice because of the reduced length of time during which natural ice was available. In Rankin Inlet, Tootoo's hometown on Hudson Bay in Nunavut, a community of 2400 residents installed artificial ice during the summer of 2003 after eight years of lobbying for the funds (Ice is Scarce 2003).

Hockey season on natural ice, which formerly ran from September until May in the 1970s, often now begins around Christmas and ends in March, according to Jim MacDonald, president of Rankin Inlet Minor Hockey. "It's giving us about three months of hockey. Once we finally get going, it's time to stop. At the beginning of our season, we're playing [lower latitude] teams that have already been on the [artificial] ice for two or three months," MacDonald said (Ice is Scarce 2003). According to Tom Thompson, president of Hockey Nunavut, there are about two-dozen natural ice rinks in tiny communities throughout the territory, but only two with artificial ice. Both are in Iqaluit (Ice is Scarce 2003).

"In my lifetime I will not be surprised if we see a year where Hudson Bay doesn't freeze over completely," said Jay Anderson of Environment Canada. It's very dramatic. Yesterday [January 6, 2003], an alert was broadcast over the Rankin Inlet radio station warning that ice on rivers around the town is unsafe. The temperature hovered around minus 12°C. It's usually minus 37 there at this time of year" (Ice is Scarce 2003).

Along with artificial ice, some Inuit are coming to know another imported generator of greenhouse gases. During 2006, officials in Nunavut authorized the installation of air conditioners in official buildings for the first time because summertime temperatures in some southern Arctic villages have climbed into the 80s F. in recent years.

The weather is not all that has been changing in Nunavut. The very fabric of Inuit life has changed within two generations. Climate change has arrived as Inuit society changes fundamentally in many ways. Inuit elders describe the coming of *qallunaat*, the non-Inuit money economy, in natural metaphors, as a time when the old world shifted.

5.16 The Arctic Ocean Is Acidifying

The Arctic Monitoring and Assessment Program reported in 2013 that ocean water at the top of the world is not only rapidly melting; it also is acidifying due to rising levels of carbon dioxide. This change continued as part of an ongoing change for at least the next decade, with no change in direction probable until and unless the sources of acidification are removed.

The Arctic Ocean is notably vulnerable, according to Rashid Sumaila of the University of British Columbia, who was among 60 scientists who took part in the study. Aboriginal peoples living above the Arctic Circle are affected. "Aboriginal people actually depend a lot on the living sources in the Arctic," Sumaila told CBC News. "They are very connected to the system, and they will be the first ones to be hit by this" (Climate Change: Arctic 2013). While acidity in oceans worldwide has increased 30% in 200 years, levels are rising even more sharply in the Arctic Ocean because it receives more fresh water than southerly latitudes. Arctic water also is relatively cold, and therefore is less able to neutralize the increased acidity.

"Because Arctic marine food webs are relatively simple, Arctic marine ecosystems are vulnerable to change when key species are affected by external factors," the monitoring program said that "Arctic marine ecosystems are highly likely to undergo significant change due to ocean acidification." As a result, "Marine mammals, important to the culture, diets and livelihoods of Arctic Indigenous Peoples and other Arctic residents could also be indirectly affected through changing food availability," the research team said (Climate Change: Arctic 2013).

5.17 Alaskan Villages Fall to Encroaching Seas

In November 2013, Barack Obama's White House held a Tribal Nations Conference that ended with an agreement to address global warming's effects on peoples represented by the National Congress of American Indians (NCAI). During the conference, the Indian Country Media Network reported that "Several federal officials noted the severe impacts that climate change has had on American Indians and Alaska Natives" (Feds Reach Out 2013). "The health of tribal nations depends on the health of tribal lands. So it falls on all of us to protect the extraordinary beauty of those lands for future generations," Obama said at the Tribal Nations Conference. "And already, many of your lands have felt the impacts of a changing climate, including more extreme flooding and droughts" Obama told the conference (Feds Reach Out 2013).

"It is critically important that tribal leaders are at the table because too often, Native voices are left out of federal conversations around mitigating the effects of climate change," the NCAI said. "Indian country faces some of the most difficult challenges stemming from climate change because of the remote location of many tribal lands and, particularly in Alaska, the dependence on the land and animals for subsistence living. NCAI applauds the Administration for this effort and is hopeful that by working together, government-to-government, tribal communities will have the tools necessary to address climate change" (Feds Reach Out 2013).

As this conference met, several coastal Alaskan Native villages were declared disaster areas following a series of severe storms. By 2013, more than 30 Alaska Native villages were experiencing or facing eventual destruction or relocation due to climate change, including Shishmaref and Newtok. "Alaska is seeing all these things the rest of the country hasn't seen yet," said Jerome Montague, Native Affairs and Natural Resources Advisor for the Alaskan Command Joint Task Force (Jessepe 2012). A new sea wall was built during 2006 in the village of Kivalina, Alaska's, seaward side, costing $3 million, hopefully to quell erosion. Other such attempts had failed as storm-driven winds and waves surged above, around, and (in some cases) under any kind of barrier that human beings could erect.

In Kivalina, erosion and flooding poured sediment into the community water system, requiring filtration to prevent ingestion of unsanitary water. About 400 people in Kivalina could take only sponge baths, with very limited access to laundry during the winter. "As hand-washing and bathing decreased, respiratory and skin diseases increased," health aides said (Climate Change Puts 2011). In Point Hope, algae blooms aided by warming water clouded a lake used for drinking water. Kivalina hangs precariously onto a slender, eroding peninsula, as the U.S. Army Corps of Engineers projects that the sea will wipe it off the map by 2025.

And, in addition, the threat of food poisoning increases with rising temperatures. Meat stored in ice cellars carved from melting permafrost can become contaminated with pathogens that cause sickness. "We used to have frozen whale meat and maktuk [most often made from the skin and blubber of bowhead whales] all winter and summertime, too," said Joe Towksjhea, a Point Hope resident, in the consortium's report. "It is not frozen anymore" (Climate Change Puts 2011).

According to Sylvester Ayek, who hunts on King Island in the Bering Sea, the greater problem is the dwindling number of animals due to late freezes and early thaws. "What we hunt is usually around ice. And when the ice goes earlier, like these past few years, the game is gone," Ayek told Indian Country Today Media Network. This issue appears to drastically affect villagers, and especially elders, in Point Hope, leading to more reports of malnutrition and anemia. Adding to the lack of available, safe food, the period for rack-drying fish, seal, and caribou has shortened because of warming weather, increasing the likelihood of food-borne illnesses caused by bacteria—not to mention the lingering scent of raw meat in milder temperatures that attracts hungry polar bears (Climate Change Puts 2011).

The Indian Country Media Network reported that "The Yup'ik village of Kotlik, Alaska, along with Unalakleet and other predominantly Native communities, were ravaged beginning November 9 [2013] by a series of four storms that battered hundreds of miles of Alaska's west coast with near hurricane-force winds, a sea surge as high as nine feet, freezing rain, and snow," Alaska Gov. Sean Parnell said. The storm surge wrecked food supplies (Yup'ik 2013). For a time, the entire town was flooded with seawater and chunks of ice. Rising seas and coastal erosion directly threaten Tuktoyaktuk, a Dene and Inuit community located at the edge of the Arctic Ocean. Ice that once protected the coast has receded out to sea. Extensive erosion washed away the school and has forced the village to relocate many other structures from cutting the town's main road, and threatens to wash its dump out to sea.

5.18 Moving Beyond a Point of No Return

Both scientists and Native peoples were shaken by the sudden retreat of the Arctic ice during the summer of 2007, much more than either had anticipated. Ice continued to retreat after that in an irregular fashion, year by year. "The Arctic is often cited as the canary in the coal mine for climate warming," said Jay Zwally, a climate expert at NASA, told the Associated Press. "Now as a sign of climate warming the canary has died" (Kolbert 2007, 44). "We could very well be in that quick slide downward in terms of passing a tipping point," said Mark Serreze, a senior scientist at the National Snow and Ice Data Center, in Boulder, Colo. "It's tipping now. We're seeing it happen now" (As Arctic 2008). Bob Corell, who headed a multinational Arctic assessment, said: "We're moving beyond a point of no return" (As Arctic 2008).

By 2007, Baffin Island had lost half of its ice in 50 years, as glaciers on its northern mountain ranges melted. In another 50 years, what remains may be gone, according to research by geological sciences Professor Gifford Miller of the University of Colorado-Boulder's Institute of Arctic and Alpine Research and colleagues. "Even with no additional warming, our study indicates these ice caps will be gone in 50 years or less," he said (Anderson et al. 2008). In a separate study released in March, 2006, Stroeve and her team showed that dwindling Arctic sea ice may have reached "a tipping point that could trigger a cascade of climate change reaching into Earth's temperate regions" (Arctic Ice Retreating 2007). "This suggests that current model projections may in fact provide a conservative estimate of future Arctic change, and that the summer Arctic sea ice may disappear considerably earlier than IPCC projections," said Stroeve (Arctic Ice Retreating 2007). In the past, low-ice years often were followed by recovery the next year, when cold winters favored accumulation, or cool summers kept much of the ice from melting. That kind of balancing cycle stopped after 2002. "If you look at these last few years, the loss of ice we've seen, well, the decline is rather remarkable," said Serreze (Human 2004, B-2).

At the same time, Alaska's boreal forests are expanding northwards at a rate of about 100 kilometers per 1 degree C. rise in temperatures. Ice cover on lakes and rivers in the mid-to-high northern latitudes now lasts for about two weeks less than it did 150 years ago. During late June, 2004, as temperatures in Fairbanks, Alaska, peaked in the upper 80s on some days, a flood warning was issued near Juneau, Alaska—not for rainfall, but for glacial snowmelt.

Arctic ice provides an important breeding ground for plankton, NASA senior research scientist Josefino Comiso said. Plankton comprise the bottom rung of the ocean's food chain. "If the winter ice melt continues, the effect would be very profound especially for marine mammals," Comiso said. Arctic ice sometimes melts even in winter as water warms, Comiso said. The winter ice season shortens every year (Arctic Ice Melting 2008).

5.19 Salmon Decline in Warming Waters

The decline of Pacific salmon runs suggests that global climate change could devastate fish populations on which millions of Native (and other) people rely for food. In Alaska, fewer salmon have returned from the ocean to spawn at their birthplaces. Those that do return have been smaller than prior averages and have arrived later than usual. The key factor in the decline of salmon runs appears to have been the fact that water temperatures often have been much higher than usual, stressing these cold-water fish.

Salmon are very sensitive to temperature. If the temperature rises 2°C. (from a usual late-summer peak of 21 to 23°C.) none of the salmon will survive the swim upstream to spawn. At 23°C., most salmon will die of heat prostration. In a living example of this, the Pacific Northwest experienced an exceptionally hot summer in 2021, with air temperatures reaching 108 F. in Seattle and 121 F. in inland British Columbia. In Seattle, Salmon offshore died in the tens of thousands (at least) because of overheated habitat. At the same time, Inuit were observing salmon fleeing the heat by swimming northward through the Being Strait.

For Canada's western regions, climate models forecast an increase in precipitation, water runoff, and flooding in winter and a decrease in precipitation and runoff during summer. Higher winter river flows are expected to damage salmon spawning grounds, reduce survival and growth of fish because of increased stream temperatures, and damage Fraser River salmon due to increased predation by warm-water species. In 1995, the Canadian Department of Fisheries and Oceans blamed collapse of Fraser River salmon runs on predation by mackerel, which invaded the salmon spawning grounds along with warmer-than-average ocean waters provoked by El Nino conditions. At the same time, and (according to the Canadian government, for the same reasons) the Queen Charlotte Chinook Salmon runs declined about 80%. Warmer sea water decreases the population of krill which the salmon eat. Anglers also say that sea lice from salmon reared on fish farms are infesting wild salmon, often killing them. The fish farms also produce a polluting slurry. In eight of thirty-two rivers on the west coast of Scotland, salmon were virtually extinct by the year 2000. Seals are killing salmon in some rivers as well.

Rising temperatures in British Columbia's largest sockeye Salmon spawning river, the Fraser, provoked a government shutdown of commercial salmon fishing at the height of the season late in September 1998.

Between a quarter and two-thirds of salmon returning to various locations in the Fraser River system were dying before they could spawn, many of them from conditions related to rapid warming of their aquatic environment. The water warmed because of below-average snowpack and above-normal temperatures in the interior of British Columbia, the Fraser's drainage basin. A large aluminum smelter also was delivering of heated water to an upstream tributary of the Fraser, adding yet another human provocation to the salmon's warming environment. Clear-cutting of upstream forests is also blamed for some of the warming of the river's waters. By late July 2006, the snow pack in the Fraser River watershed of British Columbia, which in most years would be feeding into the river and cooling it had almost entirely melted. Salmon were threatened by a combination of warm water and low river levels—as often happens on today's Earth, by a collusion of climatic change provoked by industrial development, both attributable to human activity.

Shifts in climate, and responses to them, occur so quickly that they are easily recognizable over a short period of time. Robert Way, a climatologist based in Happy Valley-Goose Bay Area of Labrador, said that in the region, compared to historical norms, winter is about six weeks shorter, while the region's sea ice coverage is about a third smaller than it was a decade ago. If you look at the rate of warming from the late 1980s to 2015, this is one of the fastest warming places in the world. "It's quite concerning," said Way. Traditionally oriented Inuit are worried that hunting and trapping skills may become irrelevant in this new reality. More and more, Inuit are relying on expensive, store-bought processed foods because it is safer and easier than catching or shooting supper. Their ancestors never experienced a time when their frozen world in northern Labrador was being altered so dramatically because of climate change. Shrinking ice packs and more severe weather have made travel increasingly difficult and dangerous, often cutting people off from other communities and traditional hunting lands (Mercer 2018). Traditional knowledge that has been used by the Inuit for several thousand years by the Inuit is disappearing with the melting ice.

"Everything we do involves the ocean and sea ice, which is highly affected by climate change," said Natasha Simonee, a member of the Inuit community in Pond Inlet. Diminishing ice, warmer winters, and changes to weather patterns are not just background noise for people like Simonee. These things have the potential to alter her entire way of life. She explains how elders in her community have grown reluctant to share traditional methods for predicting the weather, since the climate has changed so much. "They question whether it's accurate," she explained. "That kind of stuff is not necessarily practised and passed on anymore" (Baraniuk 2021).

The measures of life in the Arctic are changing as the Inuit watch. "Summers have always been short here and marked by tormenting swarms of black flies." Paula McLean-Sheppard, a Nunatsiavut government employee, said she has been startled to see the insects arriving earlier and earlier in the spring. "I was driving my truck last year and there was this big black cloud that suddenly appeared on the windshield" she said. "I had to use my wipers to clear it, I didn't know what it was," she said. "It was the *flies*. It used to be the end of June before you saw your first fly. Now they're coming in May" (Mercer 2018).

5.20 Climate Change and Industrial Development

For a segment of Inuit people, melting of ice and general warming provides promise for economic development and profit, non-withstanding loss of culture or animals such as polar bears. Others, however, take a look at a likely future and express profound concern. The United Nations Environment Programme (UNEP) characterizes the circumpolar Arctic as the world's climate change "barometer." The 160,000 Inuit who live in northern Canada, Greenland, Alaska, and Chukotka in Russia have witnessed the changing of the natural environment as a result of global warming for almost 20 years. Far easier access, particularly by sea, will be available to the Arctic's minerals and hydrocarbons, many of which are located offshore. A significant increase in general cargo transits is projected through the northwest or northeast passages, or even the Arctic Ocean. In short, climate change will promote and accelerate industrial development in a unique, fragile, and vulnerable region. "It is not far-fetched to foresee shipping in the Arctic, linking Europe and Asia to the western and eastern seaboards of North America, cutting off thousands of miles of global sea routes, which will further impact our sensitive region" (Smith 2021).

The circumpolar Arctic may well become a region of considerable geopolitical and strategic importance. Some authors have predicted mass population movements as a result of climate change. This may be plausible in tropical and temperate regions, but it remains highly unlikely in the Arctic. Nevertheless, how will the region's indigenous populations fare in a future influenced by global climate change? Firm answers are not possible, but adaptation on a huge scale will be needed despite the risks involved. The culture of Inuit and other Arctic indigenous peoples is based on their relationship with the land, environment, and animals. Wholesale adaptation to an industrial future may be tantamount to assimilation that indigenous peoples worldwide seek to avoid (Smith 2021).

References

Anderson, Rebecca K., Gifford H. Miller, Jason P. Briner, Nathaniel A. Lifton, and Stephen B. DeVogel. "A Millennial Perspective on Arctic Warming from 14-C in Quartz and Plants

Emerging from Beneath Ice Caps." *Geophysical Research Letters* *35*:L01502; https://doi.org/10.1029/2007/GL032057, 2008. Last accessed January 5, 2008.

"Arctic Ice Retreating 30 Years Ahead of Projections". Environment News Service, April 30, 2007. http://www.ens-newswire.com/ens/apr2007/2007-04-30-04.asp. Last accessed November 5, 2007.

"As Arctic Sea Ice Melts, Experts Expect New Low". Associated Press in New York *Times*, August 28, 2008. http://www.nytimes.com/2008/08/28/science/earth/28seaice.html. Last accessed December 10, 2008.

Baraniuk, Chris. "The Inuits' Knowledge Vanishing with the Ice." BBC News, October 11, 2021. https://www.bbc.com/future/article/20211011-the-inuit-knowledge-vanishing-with-the-ice. Last accessed October 30, 2021.

Bowen, Jerry. "Dramatic Climate Change in Alaska." C.B.S. News Transcripts; C.B.S. Morning News 6:30 a.m. Eastern Daylight Time, August 29, 2002. (In LEXIS)

Carlton, Jim. "A Winter Without Walrus: Harvesting of Food Staple for Remote Eskimo Villages Plummets; Disaster Declared." *Wall Street Journal*, October 4, 2013, A-4.

"Climate Change: Arctic Ocean Is Acidifying Fast, Study Shows". Indian Country Today Media Network, May 15, 2013. http://indiancountrytodaymedianetwork.com/2013/05/15/climate-change-arctic-ocean-acidifying-fast-study-shows-149369. Last accessed July 21, 2013.

"Climate Change Puts Health of Arctic Villagers on Thin Ice". Indian Country Today Media Network, March 07, 2011. http://indiancountrytodaymedianetwork.com/article/climate-change-puts-health-of-arctic-villagers-on-thin-ice-21391. Last accessed March 20, 2011.

Climate Change Creating Vast New Glacial Lakes, with Risk of 'Gargantuan' Floods, Researcher Says. The Canadian Press. August 31, 2020. https://www.rcinet.ca/eye-on-the-arctic/2020/08/31/climate-change-creating-vast-new-glacial-lakes-with-risk-of-gargantuan-floods-researcher-says/. Last accessed February 2, 2021.

Cozzetto, K., K. Chief, K. Dittmer, M. Brubaker, R. Gough, K. Souza, F. Ettawageshik, S. Wotkyns, S. Opitz-Stapleton, S. Duren, and P. Chavan. "Climate Change Impacts on the Water Resources of American Indians and Alaska Natives in the U.S." *Climatic Change* 120(2013):569-584.

Dowsley, M, Gearheard, S, Johnson, N, Inksetter, J. 2010. Should we turn the tent? Inuit Women and Climate Change. *Inuit Stud* 34(1): 151–165. https://doi.org/10.7202/045409ar. Last accessed January 215, 2010.

Eilperin, Juliet. "Alaskan Arctic Villages Hit Hard by Climate Change." Washington *Post,* August 5, 2012. http://www.washingtonpost.com/national/health-science/alaskan-arctic-villages-hit-hard-by-climate-change/2012/08/05/e9dbd4a6-d5b0-11e1-a9e3-c5249ea531ca_print.html. Last accessed August 30, 2012.

"Feds Reach Out to Natives on Climate Change at Tribal Nations Conference." Indian Country Media Network, November 17, 2013. http://indiancountrytodaymedianetwork.com/2013/11/17/feds-reach-out-natives-climate-change-tribal-nations-conference-152290. Last accessed December 2, 2013.

Frey, Darcy, "George Divoky's Planet." *New York Times Sunday Magazine*, January 6, 2002, 26-30.

Gillis, Justin. "Sea Ice in Arctic Measured at Record Low." *New York Times,* August 27, 2012. http://www.nytimes.com/2012/08/28/science/earth/sea-ice-in-arctic-measured-at-record-low.html. Last accessed September 20, 2012.

Huet, C, Rosol, R, Egeland, GM. 2012. "The Prevalence of Food Insecurity is High and the Diet Quality Poor in Inuit Communities." *Journal of Nutrition* 142(3): 541–547. https://doi.org/10.3945/jn.111.149278. Last accessed March 15, 2013.

Human, Katy. "Disappearing Arctic Ice Chills Scientists; A University of Colorado Expert on Ice Worries that the Massive Melting will Trigger Dramatic Changes in the World's Weather." Denver *Post*, October 5, 2004, B-2.

"Ice a Scarce Commodity on Arctic Rinks: Global Warming Blamed for Shortened Hockey Season." *Financial Post* (Canada), January 7, 2003, A-3.

References

Jans, Nick. "Living With Oil: The Real Price." In Steven Kazlowski. *The Last Polar Bear: Facing the Truth of a Warming World.* Seattle: Braided River Press, 2008, 147-161.

Jessepe, Lorraine. "Alaskan Native Communities Facing Climate-Induced Relocation." Indian Country Today Media Network, June 21, 2012. http://indiancountrytodaymedianetwork.com/article/alaskan-native-communities-facing-climate-induced-relocation-119615. Last accessed June 30, 2012.

Joling, Dan. "Walruses Abandon Ice for Alaska Shore." Associated Press in *Washington Post,* October 4, 2007a. http://www.washingtonpost.com/wp-dyn/content/article/2007/10/04/AR2007100402299_pf.html. Last accessed November 4, 2007.

Joling, Dan. "Thousands of Pacific Walruses Die; Global Warming Blamed." Associated Press, December 14, 2007b, n.p.

Kerr, Richard A. "Ice-Free Arctic Sea May Be Years, Not Decades, Away." *Science* 337(September 28, 2012):1591.

Kolbert, Elizabeth. "Testing the Climate." (Talk of the Town) *The New Yorker*, December 24 and 31, 2007, 43-44.

Lee, Jaeah. "In Arctic, Warmer Climate Means More Beavers Defecating and Disease." *Mother Jones,* March 4, 2011. n.p.

Lynn, Kathy, John Daigle, Jennie Hoffman, Frank Lake, Natalie Michelle, Darren Ranco, Carson Viles, Garrit Voggesser, and Paul Williams. "The Impacts of Climate Change on Tribal Traditional Foods." *Climatic Change* 120(2013):545-556.

McGrath, Susan. "On Thin Ice." *National Geographic*, July, 2011, 64-75.

Mercer, Greg. "Sea, Ice, Snow … It's all Changing:' Inuit Struggle with Warming World." *The Guardian*, May 30, 2018. https://www.theguardian.com/world/2018/may/30/canada-inuits-climate-change-impact-global-warming-melting-ice. Last accessed March 20, 2022

"Nunavut Capital Battles Heat Wave". Indian Country Today Media Network January 5, 2011. http://indiancountrytodaymedianetwork.com/article/nunavut-capital-battles-heat-wave-9443. Last accessed January 3, 2011.

Qiu, Jane. "Arctic Ecology: Tundra's Burning." *Nature* 461(September 3, 2009):34-36.

Pegg, J.R. "The Earth is Melting, Arctic Native Leader Warns." Environment News Service, September 16, 2004., n.p.

"Rare Sighting Of Wasp North Of Arctic Circle Puzzles Residents". Canadian Broadcasting Corporation, September 9, 2004. http://www.cbc.ca/story/science/national/2004/09/09/wasp040909.html. Last accessed October 23, 2004.

Sansoulet, J, Therrien, M, Delgove, J, Pouxviel, G, Desriac, J, Sardet, N, Vanderlinden J. 2020. "An Update on Inuit Perceptions of Their Changing Environment, Qikiqtaaluk (Baffin Island, Nunavut). *Elementa: Science of the Anthropocene* (8:2020) https://doi.org/10.1525/elementa.025. Last accessed November 3, 2020.

Simmonds, Ian and Kevin Keay. "Extraordinary September Arctic Sea Ice Reductions and Their Relationships with Storm Behavior over 1979–2008, *Geophysical Research Letters* 36(October 14, 2009) L19715, https://doi.org/10.1029/2009GL039810. Last accessed November 1, 2009.

Smith, Duane. "Climate Change in the Arctic: An Inuit Reality." *U.N. Chronicle:* New York: United Nations. Last accessed December 10, 2021. https://www.un.org/en/chronicle/article/climate-change-arctic-inuit-reality. Last accessed January 16, 2022.

Therrien, M. 2007. Les Inuits ne trouvent pas cela extrême. 639e conférence de l'Université de tous les savoirs (UTLS), June 28, 2007. Available at https://www.canal-u.tv/video/universite_de_tous_les_savoirs/les_inuits_ne_trouvent_pas_ca_extreme_michele_therrien.1518. Last accessed June 28, 2007.

Tingley, Martin P. and Peter Huybers."Recent Temperature Extremes at High Northern Latitudes Unprecedented in the Past 600 Years." *Nature* 496(April 11, 2013):201-205.

Watt-Cloutier, Sheila. Speech to Conference of Parties (COP) to the United Nations Framework Convention on Climate Change. Milan, Italy, December 10, 2003.

Watt-Cloutier, Sheila. Personal Communication, from Iqaluit, Nunavut, January 4, 2004.

Watt-Cloutier, Sheila. "It's Time to Listen to the Inuit on Climate Change." *Canadian Geographic*, November 15, 2018. https://www.canadiangeographic.ca/article/its-time-listen-inuit-climate-change. Last accessed December 14, 2018.

Weatherhead, E, Gearheard, S, Barry, RG. 2010. Changes in Weather Persistence: Insight from Inuit Knowledge. *Global Environ Chang* 20: 523–528. https://doi.org/10.1016/j.gloenvcha.2010.02.002.

Windeyer, Chris. "South Baffin Swelters in Winter Heat Wave; 'It doesn't show any signs of abating.'" *Nunatsiaq News,* January 4, 2011. http://www.nunatsiaqonline.ca/stories/article/98789_south_baffin_swelters_in_winter_heat_wave/. Last accessed February 15, 2012.

Wohlforth, Charles. *The Whale and the Supercomputer: On the Northern Front of Climate Change*. New York: North Point Press/Farrar, Strauss and Giroux, 2004.

"Yup'ik Villages Ravaged by Fierce Alaska Storms". Indian Country Media Network, November 20, 2013. http://indiancountrytodaymedianetwork.com/2013/11/20/yupik-villages-ravaged-fierce-alaska-storms-152341. Last accessed November 28, 2013.

Chapter 6
The Inuit (and Others): If it Swims, It's Probably Poisonous

The Arctic, in today's world, is nowhere nearly as pristine as many people believe. In Nunavut, the semi-sovereign Inuit Nation in northern Canada, it has been snowing poison, as polychlorinated biphenyls (PCBs, also known as dioxins) and other chemicals produced by southern industries are swept northward by prevailing winds. At Akwesasne, in northernmost New York State and nearby Canada, the fish, laced with the same chemicals, are often inedible, and mothers also have been told not to breastfeed babies (and to avoid eating anything else from the St. Lawrence River). The Huicholes of Mexico also live in a land laced with toxic chemicals. In all of these locations, Native peoples have organized to restore a livable environment. Sometimes they have succeeded on this long and torturous road.

The Inuit, in particular, have taken an active role in efforts to outlaw the use of these toxins in an international agreement, the Stockholm Convention. Following negotiation of the Stockholm Convention, which outlaws most of the "Dirty Dozen" persistent organic pollutants (POPs), Inuit activist Sheila Watt-Cloutier evoked tears from some delegates with her note of gratitude on behalf of the Inuit people. The treaty, she said, had "brought us an important step closer to fulfilling the basic human right of every person to live in a world free of toxic contamination. For Inuit and [other] indigenous peoples, this may mean not only a healthy and secure environment, but also the survival of a people. For that I am grateful. *Nakurmuk*. Thank you" (Cone 2005, 202).

It's a long road back, however. These are nightmarish chemicals, concocted by scientists who thought they had beneficial reasons for exposing healthy people to such brutal, lethal inventions. Once they become part of the food chain, POPs are very difficult to dislodge from the bodies of human beings and other animals. Following the negotiation of international law to outlaw these pollutants came a decades-long battle to enforce the ban and to rid the Arctic food chain of their effects. POPs tend to become locked in the ecosystem, and as of this writing, Inuit still consume traditional foods and breastfeed their babies with caution.

In Inuit country, what new studies I have seen indicate that things are now worse than 10–20 years ago. Human beings can ban these chemicals, but the PCBs, et al.

are not listening. They bio-accumulate up the food chain, getting more potent in geometric progression as they go, and they never "die." As with nuclear weapons, such ingenuity concocted by humanity's ill-informed impulses has been very dangerous.

Travel nearly anywhere in North America, and toxicity often has become nearly endemic. In Alaska, the Salt Chuck Mine, a source of copper, gold, palladium, and silver between 1916 and 1941, contaminated the Kasaan, Alaska, harvesting grounds for fish, clams, cockles, crab, and shrimp. For decades, the Native peoples there and elsewhere were unaware that their harvests were saturated with effluvia from mine tailings. Even after the area was declared a Superfund site, Pure Nickel, Inc., sought to reactivate mining in 2012.

In California, the Elem Band of Pomo Indians now are suffering elevated levels of mercury in their bodies due to the Sulfur Bank Mine, a Superfund site that adjoins their colony. Hansen (2014) wrote that "Nearby Clear Lake is the most mercury-polluted lake in the world, despite the E.P.A.'s spending about $40 million over two decades trying to keep mercury contamination out of the water. Although the E.P.A. has cleaned some of the soil under Pomo roads and houses, pollution still seeps beneath the earthen dam built by the former mine operator, Bradley Mining Co. For years, Bradley Mining has fought the government's efforts to recoup cleanup costs."

6.1 "We Feel Like an Endangered Species"

"As we put our babies to our breasts, we are feeding them a noxious, toxic cocktail," said Sheila Watt-Cloutier, a grandmother who also has served as president of the Inuit Circumpolar Conference. "When women have to think twice about breast-feeding their babies, surely that must be a wake-up call to the world" (Johansen 2000, 27). Watt-Cloutier was raised in an Inuit community in northern Quebec. Unknown to her at that time, toxic chemicals were being absorbed by her body and by those of other Inuit in the Arctic.

During a career as activist and diplomat, Watt-Cloutier ranged between her home in Iqaluit (pronounced "Eehalooeet"), capital of the new semi-sovereign Nunavut Territory, to and from Ottawa, Montreal, New York City, and other points south, doing her best to alert the world to toxic poisoning and other perils faced by her people. The Inuit Circumpolar Conference represents the interests of roughly 140,000 Inuit spread around the North Pole from Nunavut (which means "our home" in the Inuktitut language) to Alaska and Russia. Nunavut itself, a territory four times the size of France, has a population of roughly 27,000, 85% of whom are Inuit.

Many people who live the temperate zones share stereotypes of an Arctic as a pristine place largely without many of the human pollutants that are so ubiquitous there. If you are a tourist lacking interest in environmental toxicology, today's Inuit homeland can *seem* wonderfully pristine during its long, snow-swept winters. The

Inuit still hunt seals and polar bears on the pack ice. The real problems are invisible, however. It's what you *can't* see that may kill you. The scenery may *look* pristine until one realizes the fat of polar bears, seals, and other warm-blooded animals are laced with PCBs and dioxins.

6.2 A Poisoned World

Roald Amundsen, an early European explorer of the Earth's polar regions in 1908 said that peoples living there "living absolutely isolated from civilization of any kind are undoubtedly the happiest, healthiest, most honorable and most contented of peoples.... My sincerest wishes for our Nechilli Eskimo friends is that civilization may never reach them" (Amundsen 1908, 93).

Little more than one century after he made this statement, environmental threats that Inuit face today would sound like science fiction. Welcome to a land which is on the road to environmental apocalypse: a people who never asked for any of what industries southward have burdened them with. Persistent organic pollutants such as DDT, dioxins, and PCBs have been multiplying into their food chains as air and ocean currents carry the poisonous emissions from lower latitudes to their homelands. The Arctic, surrounded by a ring of industry from Russia, to Europe, to North America, is absorbing poisons which lodge in Inuits' bodies and biomagnify, becoming more potent with each passing generation, cursing the people with an environmental curse.

"We are the miner's canary," said Watt-Cloutier. "It is only a matter of time until everybody will be poisoned by the pollutants that we are creating in this world" (Lamb n.d.). "At times," said Cloutier, "We feel like an endangered species. Our resilience and Inuit spirit and of course the wisdom of this great land that we work so hard to protect gives us back the energy to keep going" (personal communication, March 28, 2001).

Nearly all of the chemicals that now afflict Inuit are synthetic compounds of chlorine. Some of them are very toxic. One-millionth of a gram of dioxin has the capacity to kill a guinea pig, for example. Nearly all of the flora and fauna of the Arctic have been polluted by these chemicals, and others, including high levels of mercury, lead, and nuclear radiation.

6.3 How Persistent Organic Pollutants (POPs) Work

The POPs are causal factors in many birth defects, cancers, and other immune system, reproductive, and neurological damage to humans and other animals. At higher levels, the same chemicals also can damage the central nervous system. Many of the same chemicals also produce endocrine disruptions and cause deformities dysfunction of reproductive systems. These POPs also may produce Cancers of the brain and endocrine system by penetrating the placental barrier and thereby misdirecting instructions of a body's natural chemical messengers. These influence development of a fetus by telling it how to develop from before birth, in the womb through puberty. Once such interference take place, nervous, immune, and reproductive systems often do not develop as programmed by genes that have been inherited by the embryo.

These poisons can poison people and other animals years after they have been released by industries to the south. Although the United States is several hundreds of miles from the Arctic, as much as 82% of persistent organic pollutions that Nunavut receives up to 82% of its dioxins from industries there (Suzuki 2000). Some pollution in the Canadian Arctic has been emitted in Western Europe and Japan. Contamination reaches the Arctic from Europe in about two weeks.

The cold climate of the Arctic also decreases decomposition of these substances. Thus, they persist longer than at lower latitudes. The Arctic acts as a "cold trap," collecting many industrial pollutants, including toxaphene, chlordane, PCBs, and mercury. As a result, "Many Inuit have levels of PCBs, DDTs and other persistent organic pollutants in their blood and fatty tissues that are five to ten times greater than the national average in Canada or the United States" (PCB Working Group n.d.).

The generation of POPs has become an issue in Watt-Cloutier's home, Iqaluit, on Baffin Island, where the town dump burns wastes that emit dioxins and furans. The dump's plume provides only a small fraction of the Iqaluit residents' exposure to POPs, but it has become enough of an issue to figure in a three-month shutdown of the dump that caused garbage to pile up in the town. The dump was reopened after local public health authorities warned that the backlogged garbage could spread disease and that "the hazard posed by the rotting piles of garbage outweighed the risks of burning it" (Hill 2001, 5). In 2001, Iqaluit's government was asking residents to separate plastics and metals from garbage that can be burned without adding POPs to the atmosphere.

These chemicals accumulate in each succeeding generation in breast-feeding mammals, including the Inuit as well as a majority of their land and sea food sources. The mammalian reproductive cycle compounds their toxic effects. These toxic substances often are absorbed by plankton through airborne transmission. Small fish are then eaten by whales, dolphins, as well as other large mammals, whose large amounts of fat stores that carry the hazardous substances. They are then transmitted to offspring through breast-feeding. This kind of toxicity is more vulnerable to sea mammals than land animals, so that levels of hazardous chemicals in their bodies can reach very high levels. These POPs "bioaccumulate" in mammals at the top of their respective food chains—in the bodies of large meat eaters, such as marine mammals, polar bears, raptors, and humans. As will be described below, breast-feeding of a human infant further raises these levels. Whales, seals, and dolphins, in the northern seas also are being contaminated. Large land animals, including caribou also are affected. Humans who eat these animals are the top (most polluted) rung on the food chain.

6.4 Inuit Infants: "A Living Test Tube for Immunologists"

Inuit infants have provided "a living test tube for immunologists" (Cone 1996, A-1). Inuit women's breast milk contains six to seven times the PCB levels of women in urban Quebec because they eat contaminated sea animals and fish, according to Quebec provincial government. Their babies also have had very high rates of bronchitis, pneumonia, meningitis, and other infections relative to other Canadians. One Inuit child of four, on average, has chronic hearing loss because of infections. "In our studies, there was a marked increase in the incidence of infectious disease among breast-fed babies exposed to a high concentration of contaminants," said Eric Dewailly, a Quebec Public Health Center researcher who was one of the first scientists who, during the mid-1980s, found very high levels of POPs in Inuit mothers' breast milk.

Dewailly, a Laval University scientist at the time, accidentally discovered that, due to their intake of PCBs from their mothers' milk, Inuit infants were being heavily contaminated by PCBs. Dewailly had initially visited the Inuit seeking a test group without POP *without* such contamination to use as a baseline to compare with women in southern Quebec who had tested positive for PCBs in their breast milk. Instead, Dewailly discovered that Inuit mothers' PCB levels that were at several times those of Quebec mothers. Until that time, Arctic sea mammals had not been suspected.

Dewailly and colleagues then switched the focus of their studies to the Arctic (Dewailly et al. 1993, 1994, 2000), and from mothers to infants, seeking evidence whether exposure to organochlorines could be implicated as a source for POP-influenced incidences of infectious diseases in Inuit infants from Nunavut. Dewailly and his colleagues later reported that serious ear infections among infants were at least twice as common among Inuit babies whose mothers tested higher-than-usual concentrations of toxic chemicals in their breast milk. More than 80% of 118 babies studied in various Nunavut communities had at least one serious ear infection in the first year of life (Calamai 2000).

The three most common contaminants that researchers found in Inuit mothers' breast milk were three pesticides (dieldrin, mirex, and DDE, a derivative of DDT) as well as two industrial chemicals: PCBs and hexachlorobenzene. At first, the researchers could not determine which specific chemicals made Inuit babies more vulnerable, or whether effects of these chemicals intensify in combination.

The Arctic Monitoring and Assessment Program, jointly operated by Arctic nations and organizations of other indigenous Arctic peoples, in its study *Pollution and Human Health* that "PCB blood levels, while highest in Greenland and the eastern Canadian Arctic, were high enough (over 4 micrograms of PCBs per liter of blood) that a proportion of the population would be in a risk range for fetal and childhood development problems" (PCB Working Group n.d.). Because they are not easily broken down or excreted, these compounds remain in the body for months or years.

Because they often are born with depleted white blood cells, Inuit children suffer many more bouts with diseases than others, including a 20-fold increase in life-threatening meningitis compared to other Canadian children. These childrens' immune systems are so depleted that they do not produce enough antibodies to resist even common childhood diseases. Inuit babies also have experienced strikingly high rates of bronchitis, pneumonia, and other infections compared with other Canadians. One Inuit child of four, on average, experiences chronic hearing loss due to infections. A study published in the September 12, 1996, issue of the *New England Journal of Medicine* confirmed that children exposed to low levels of PCBs in the womb grow up with lower-than-average IQs, memory problems, difficulty paying attention, and poor reading comprehension.

According to the Quebec Public Health Center, 1052 parts per billion (ppb) of PCBs that have been found in Arctic women's milk fat. This compares to a reading of 7002 ppb in some polar bear fat, 1002 ppb in whale blubber, 527 ppb in seal blubber, and 152 ppb in fish. By contrast, the U.S. Environmental Protection Agency (EPA) safety standard for edible poultry is 3 ppb. For fish, it is 2 ppb. At 50 ppb, soil is often classified as hazardous waste. Research by the Canadian Federal Department of Indian and Northern Affairs indicated that Inuit women throughout Nunavut have DDT levels nine times that of averages in some Canadian urban areas. Milk of Inuit women of the eastern Arctic has been measured to contain as much as 1210 ppb of DDT and its derivative DDE, while milk from women living in southern Canada contains about 170 ppb.

Infants in the North are being developmentally harmed by exposure to several contaminants, among them lead, methylmercury, and PCBs. While IQ tests have long been criticized as culturally biased (favoring white, middle- or upper-class Anglo-American children), indigenous infants' cognitive abilities also are damaged by environmental contamination, beginning in the womb and continuing after birth. Physiological damage due to chemical exposure impedes both biological and psychological development. Each chemical has different effects.

Babies in Arctic Canada are at risk of specific effects on their mental abilities, depending on which contaminants they absorb in the womb, according to a study released in 2014. While all of the POP contaminants have been linked to neurological effects, each may have different effects on infants. For example, according to the study, "PCBs seemed to impair the babies' ability to recognize things they have seen" ("Contaminants" 2014). A study (Boucher et al. 2014) that involved 94 Inuit infants and their mothers from Nunavik in northern Quebec measured prenatal exposure to lead, methylmercury, and PCBs by testing cord blood, followed by standard mental development tests of the babies at 6.5 and 11 months of age, taking part in such activities as retrieving toys from under cloth covers, recognizing photographs, and motor-skills tasks, to identify how growing brains were being damaged. According to Boucher et al. (2014), "Each contaminant was independently associated with impairment of distinct aspects of cognitive function with long-term implications for cognitive development—PCBs with visual recognition memory, methylmercury with working memory and an early precursor of executive function, lead with processing speed—deficits that can already be detected during the first year of life."

Many of the same scientists have been building their case regarding the effects of contaminants on IQs for more than two decades. This trend is worldwide: "Similar findings have been reported for children in Michigan, Oswego, N.Y., the Faroe Islands, and Taiwan—all areas where many babies are highly exposed to PCBs or mercury from their mothers' consumption of fish or marine mammals" ("Contaminants," 2014). "The last thing we need at this time is worry about the very country food that nourishes us, spiritually and emotionally, poisoning us," Watt-Cloutier said. "This is not just about contaminants on our plate. This is a whole way of being, a whole cultural heritage that is at stake here for us" (Mofina 2000, A-12). "The process of hunting and fishing, followed by the sharing of food—the communal partaking of animals—is a time-honored ritual that binds us together and links us with our ancestors," said Watt-Cloutier (PCB Working Group n.d.).

In many cases, the Inuit have no practical alternative to "country food." In addition, it is traditional. Even though grocery stores do business in Canadian Inuit towns and villages, all grocery food is flown in, with various effects on freshness and taste. No roads, highways, or natural land bridges lead south from the villages to Canadian cities that might supply food.. The cost of air freight, compounded by distance, raises the price of a quart of milk in the high Arctic to $15 or more, and a battered head of lettuce to $8 (at 2022 prices). A tiny frozen turkey the size of a stewing hen may cost $50 or more.

6.5 Socioeconomic Changes in Nunavut

The per capita annual income in Nunavut was about U.S. $12,000 in 2022, which does not buy much in a place where two liters of milk costs $12 and a loaf of Wonder Bread retails for $7. While Nunavut has only a few paved roads and no railroads (communities are connected by airline services), all villages have television and radio; most have telephone and the Internet, so images imported from the outside world are never far away.

Widespread poverty, unemployment, crime, substance abuse, and a high suicide rate continue to beset Nunavut, which also faces high unemployment. Iqaluit (Inuktitut for "fishing place"), Nunavut about 1600 miles north of New York City), contained about 15,000 people in 2020, many of whom have big-city problems, including alcoholism, HIV-AIDS, drug abuse, and COVID-19. The Royal Canadian Mounted Police detachment at Iqaluit complains that its Mounties spend so much time rounding up drunks, so much that "it's getting in the way of fighting crime" ("Iqaluit Drums" 2001). Three thousand visits to the city's drunk tank is not unusual in a year.

The traditional (usually nomadic) Inuit hunting economy endured largely intact until the 1950s and the 1960s, when Canadian authorities compelled many people to settle in permanent communities. Many Inuit children were pressured to attend church-run boarding schools away from their families. The schools were designed to assimilate Inuit children into the culture of the *qablunaaq*, the white man. The nearly total collapse of the hunting economy has coincided with a sharp rise in suicides. The young men now often have no place to learn vital traditional skills as they did when food was acquired on the hunt. The traditional role of old men teaching skills to younger men has broken down. The young men have no place to expend their energies, or to define status.

Less than 50 years ago, suicide was nearly unknown among the Inuit. Suicide now is one of the leading causes of death there. Among young people, suicide has become by far *the* leading cause of death. Geela Patterson, an adviser for the Inuit government on women's issues and a suicide-prevention counselor, killed herself during June of 1999. Even in a town with the highest suicide rate in Canada (seven times the Canadian average), where few families have not been troubled by suicide, Patterson's death sent shock waves through the community.

Most of Nunavut's suicides have afflicted healthy young men aged 15–29. Another factor has been implicated on top of the Inuits' loss of social structure. That is the destruction of the fur markets. Anti-fur campaigns in Europe, the United States, and Canada were sustained by efforts to save the lives of fur-bearing animals. However, the same campaigns destroyed the lives of many Inuit men.

Unemployment rates in many northern villages rose to more than 50%. "When you reduce the usefulness of men in society, it is bound to have psychological effects," said Simona Arnatsiaq, leader of a Baffin Island women's organization. Susan Aglukark, an Inuit singer who has gained pop-star status in southern Canada, sings of the pain caused by suicide in ballads in English and in her Native Inuktitut. "We are investing more money for suicide prevention," said Paul Okalik, whose brother committed suicide. Rachel Attituq Qitsualik, writing in the *Nunatsiaq News*, a weekly newspaper serving the eastern Arctic, said, "I myself, when younger and miserable, have heard the siren call of suicide as a release from suffering, as has my husband, and others that I know and love.... My brother gave in to it" (Brooke 2000).

6.6 Toxic Pollution of Inuit Land by the Military

As the Inuit were being poisoned by toxins from the sky and water, they also discovered that some parts of their homelands were soiled with poisoned "hot spots" from abandoned military bases and mines. Several of these hot spots were located near the 63 radar sites in Greenland, Alaska, and Canada, that comprise the DEW (Distant Early Warning) (DEW) system. According to the Arctic Monitoring and Assessment Program's report *Arctic Pollution Issues: A State of the Arctic Environment*, these sites contained about 30 tons of PCBs, and "an unknown amount has ended up in their landfills" (PCB Working Group n.d.).

Under an agreement reached in 1998, Canadian taxpayers paid almost $720 million to clean up 51 decommissioned U.S. military sites across Canada. Cleanup of cancer-causing PCBs, mercury, various petroleum by-products, as well as lead, and radioactive materials was expected to take nearly 30 years. Under the arrangement, the United States was absolved of legal responsibility for environmental damage in Canada in exchange for $150 million in U.S. weapons and other military equipment. The cleanup of all 51 American military sites revealed pollution that newspaper reports in Canada characterized as "staggering" (Pugliese 2001, B-1). For example, at Argentia, Newfoundland, 70 miles southeast of St. John's, a U.S. Navy base that had operated from 1941 to 1994 left behind PCBs, heavy metals, and asbestos, as well as landfills filled with several other types of hazardous wastes. Waste fuels also have contaminated the water table in the area.

Abandoned DEW (Distant Early Warning) sites in the Arctic had been contaminated with discarded batteries, antifreeze agents, solvents, paint thinners, PCBs, and lead. According to news accounts, "[Canadian] Defence Department scientists have established that PCBs have leaked from the DEW line sites into surrounding areas as far away as 20 kilometers and, in some cases, the chemicals have been absorbed by plant and animal life" (Pugliese 2001, B-1). Many of the DEW line locations were established in areas where Native peoples hunt and fish. The Alaska Community Action on Toxics Program worked with indigenous communities that face contamination from Cold War sites, including the Yu'pik community on St. Lawrence Island. Alaska Community Action on Toxics also has provided the first comprehensive map of more than 2000 hazardous waste sites in Alaska, some of them owned by Native peoples.

Inuit territories also endure pollution from European and Asian pollution that flows into the Arctic Ocean. Pesticide residues and other pollutants spill into the Arctic Ocean from north-flowing rivers. Decaying Russian nuclear submarine installations on the White Sea also have polluted the ocean with nuclear wastes, including (but not limited to) entire reactor cores from scrapped ships. While many of the former Soviet Union's worst polluters have gone out of business, some prosper despite the fact that their effluent is adding to the Canadian Arctic's toxic overload. The worst offender is the Norilsk nickel smelter, located in northern Siberia. Traces of heavy metals from Norilsk's industries have been detected in the breast milk of Inuit mothers.

Geoffrey York, a reporter for the *Toronto Globe and Mail*, described the industrial city of Norilsk, population 230,000, as "the world's most polluted Arctic metropolis" (York 2001, A-1):

> Looming at the end of the road is a horizon of massive smokestacks, leaking pipes, rusting metal, gigantic slag heaps, drifting smog, and thousands of denuded trees as lifeless as blackened matchsticks. Inside malodorous smelters, Russian workers wear respirators as they trudge through the hot suffocating air, heavy with clouds of dust and gases.... Soviet[-era] research in 1988 found that Norilsk Nickel had created a 200-kilometer corridor of dead forests to the southeast of the city. (A-11).

"We, the Inuit, who for a millennium have lived in harmony with, and with great respect for our land and wildlife are now most impacted by outside forces such as POPs and climate change," said Sheila Watt-Cloutier, who was speaking for the Inuit Circumpolar Conference when she made this statement "With so much already on our plate in terms of attempting to reclaim and was speaking when she made this statement. "We must restructure our lives to gain back some control, be it personal, family, institutions, governance systems, etc. ... It can at times be overwhelming" (Watt-Cloutier, personal communication, March 28, 2001).

6.7 Toxic Contamination of Traditional Inuit Foods in Alaska and Russia

Toxic contamination of traditional foods has become an issue in Alaska as well as in Nunavut. Native delegates told a conference on Arctic pollution in Anchorage on May 1, 2000, that some Alaska natives have been avoiding their traditional foods out of fear that wild fish and game species on which they depend for food have been contaminated with heavy metals, pesticides, and other toxins.

"I have a son who has quit eating seal meat altogether," said St. Paul Island resident Mike Zacharof, president of the Aleut International Association. The association was formed because indigenous people in western Alaska and Russia are worried about pollution that crosses their countries' boundaries, Zacharof said, adding, "Fear that seal livers may contain mercury has made many islanders wary of eating the staple of their diet, though most still do" (Dobbyn 2000). "Many people in Prince William Sound no longer eat their traditional foods because of the [*Exxon Valdez*] oil spill. This impacts not only our physical well-being but our emotional and spiritual lives as well," said Patricia Cochran, director of the Alaska Native Science Commission. Cochran added, "Natives from every region of Alaska have been noticing more tumors, lesions, spots and sores on land and sea animals" (Dobbyn 2000). Indigenous reports are being compiled in a report for the Alaska Native Science Commission called the Traditional Knowledge and Contaminant Project, started in 1997, funded by the United States Environmental Protection Agency (EPA).

Research by the Circumpolar Arctic Monitoring and Assessment Program has revealed high levels of DDT in the breast milk of Russian mothers in Arctic regions, just as DDT and other toxins have been detected in Inuits' mothers' breast milk. In another study, at the University of Alaska Anchorage, indicated that pregnant Alaska Native women who eat subsistence foods may be exposing their fetuses to potentially dangerous pollutants. "There's no question that people are concerned not only about what it's going to do to them but to their unborn children," Cochran said (Dobbyn 2000). The Tanana Chiefs' Conference, a tribal social-services agency in Alaska's interior has detected high levels of DDT in salmon.

Because of toxic contamination, Alaskan Inupiat hunters now examine animals they are preparing to butcher more closely. It is called "playing doctor," wrote David Hulen, who described one hunter as he "slips on a pair of rubber gloves, wipes clean a titanium-blade knife and begins cutting tissue samples from a seal to be sent off and tested for PCBs, DDT and nearly 50 other industrial and agricultural pollutants" (Hulen 1994, A-5). The hunters, in this case, were accompanied by Paul Becker, a Maryland-based scientist from the National Marine Fisheries Service. Becker worked with native hunters, collecting samples for a national [United States] marine mammal tissue bank. He also trained several groups of villagers in Nome, Barrow, and other Alaskan settlements to take draw samples as part of their regular hunts. Hulen (1994, A-5) reported that "all of this has caused no small amount of anxiety in the villages, where oil from rendered seal fat is a staple consumed like salad dressing and where people routinely dine on dishes such as dried seal meat, walrus heart and whale steaks."

Charlie Johnson, executive director of the Eskimo Walrus Commission, a Native group that works with the government to help manage Pacific walrus populations, said, "We've seen a huge increase in cancer rates. People hear about these things turning up and wonder if that has anything to do with it. I don't think there's enough known yet to say there's a problem here in Alaska ... but I think we have to be aware that we could have a problem eventually" (Hulen 1994, A-5). These words have been prescient. Because of their role in handling bodily wastes, the kidneys of mammals are especially vulnerable to accumulation of toxic chemicals. Natives continue to eat the kidneys of the bowhead whale, however, because the kidney is one of the best-tasting parts of the animal.

6.8 Seward, Alaska: Don't Eat the Reindeer

Alaska's Seward Peninsula has been extensively mined during the past 100 years for cadmium and various lead-bearing ores that are easily absorbed by plants which provide food eaten by ungulates that concentrate in liver, kidney, and muscle tissues. Health officials have warned local native peoples to avoid eating reindeer, once a dietary staple. Contamination from weapons testing, accidental pollution, or illegal dumping also has found its way may have found its way into the lichens of northwestern Alaska, thereby accumulating in reindeer and caribou tissue.

The people on the Seward Peninsula have long lived in a subsistence lifestyle in which a high percentage of their diet comes from local plants and animals. The incidence of cancer and other diseases appears to be rising among the indigenous people in this area who subsist on contaminated reindeer and other "country food." Many people in local villages are particularly concerned that contamination from mining operations, air pollution, and dump sites are concentrating in the tissues of subsistence animals, creating health risks. The University of Alaska's Reindeer Research Program has been detecting high levels of cadmium and lead in several species. Similar concentrations are found in meat; 40 to 60 grams per week could exceed the government-recommended intake rate ("Heavy Metal" 2000).

6.9 Deformed Babies, Mercury, and the Grassy Narrows First Nation in Ontario

Deformed babies are being born on the Grassy Narrows First Nation in Ontario between 1962 and 1970, after the Dryden Chemical Co. dumped roughly 10 tons of neurotoxins into the English-Wabigoon River. More than half of the 160 people who were examined there showed symptoms of mercury poisoning, 40% of whom had Minamata disease, a type of long-term, acute, mercury contamination which produces tunnel vision, loss of coordination, seizures, and tremors.

"Both the federal and provincial governments need to recognize and effectively address the lasting issue of mercury exposure in First Nation communities along the English Wabigoon River system," Assembly of First Nations National Chief Shawn A-in-chut Atleo said. "The federal and provincial governments' callous disregard for the health and well-being of the Grassy Narrows community is appalling," said Assembly of First Nations Regional Chief Angus Toulouse, speaking for the Chiefs of Ontario, in a statement. "The poverty and ill health currently being experienced by Grassy Narrows' citizens are the direct consequences of unregulated development, and disregarding the community's right to free, prior and informed consent on activities that occur within their traditional territory" ("Mercury Poisoning," 2012).

6.10 Too Few Infant Boys? Blame Estrogen-Blocking Chemicals

The mothers of Canada's Aamjiwnaang First Nations community gave birth to a very low number of boys, with exposure to estrogen-blocking chemicals as the probable reason. "While we're far from a conclusive statement, the kinds of health problems they experience—neurodevelopment, skewed sex ratios—are the health effects we would expect from such chemicals and metals," said Niladri Basu, an associate professor at Montreal's McGill University, a lead author of a study on the matter, released in 2005 revealing that only about 35% of Aamjiwnaang infants were male (Bienkowski 2013).

The Aamjiwnaang reserve lies about 15 miles from "Chemical Valley", with more than 50 chemical factories, oil refineries, and other pollution sources, which is located near the U.S.–Canadian border near Lake Huron, east of Michigan, across the St. Clair River, where about 850 of 2,000 tribal members live. Four types of PCBs at levels two to seven times higher than average among other Canadians were measured in 42 children and mothers at Aamjiwnaang. The PCBs, which skew sex ratios, are common in fish in that area, but few of the people had eaten them. Basu suspected that the chemicals remained in the soil and air from decades ago, before they were banned in the 1970s. "Aamjiwnaang" means "at the spawning stream," describing their historical diet of fish, but anyone has been warned since the 1970s not to eat fish from the nearby lake and rivers.

"I'm struck by the elevated PCB levels," said Nancy Langston, a professor of environmental history at Michigan Technological University who was not involved in the study, "and the fact that all are anti-estrogenic," meaning that they block the hormone, while others are estrogenic, which means that they mimic it (Bienkowski 2013).

Other health problems are common in the community:

> In addition to PCBs, the mothers and children had elevated levels of cadmium, some fluorinated chemicals and the pesticides hexachlorocyclohexane and DDT compared with the Canadian average.... In addition to the skewed sex ratio, 23% of Aamjiwnaang children have learning or behavioral difficulties—a rate about six times higher than children in a neighboring county, according to a 2005 community study.... The asthma rate for children on the reservation is about 2.5 times higher than the rest of the county, according to a 2007 study by Ecojustice, a Canadian environmental organization. Birth complications also are commonplace. Of 132 women surveyed in the community in 2005, 39% had at least one stillbirth or miscarriage. The average for U.S. women is 15%, according to the National Institutes of Health.

Industrialization in the area has been increasing, with the opening of the Sarnia oil refinery, owned by Suncor Energy, has expanded to handle oil sands (a.k.a. "tar sands") production from Alberta. This is also true in other areas experiencing increases in petrochemical refining: "Men in Fort Saskatchewan—downwind of refineries and chemical manufacturers and oil sands processors—suffer from leukemia and non-Hodgkins lymphoma at higher rates than neighboring communities.... Two known carcinogens—butadiene and benzene—were found at higher levels in rural Fort Saskatchewan than in many of the world's most polluted cities," reported Isobel Simpson, lead author of the study and a chemist at the University of California, Irvine (Bienkowski 2013).

6.11 The Ouje-Bougoumou Cree Endure Severe Metal Contamination

The Ouje-Bougoumou, a Cree community in northern Quebec about 350 miles north of Montreal, has been experiencing severe metal contamination of its food and water, due to pollution related to logging and mining. During the early 1990s, the

Canadian federal government created a new community for the Ouje-Bougoumou Cree Nation and forced its members to relocate. The move has not solved the pollution's effects on fish, their main food source.

"I was kind of shocked," said Cree Margo Miascum Cooper. She fears that the water pollutants killed her father, Albert Cooper, who died of cancer in 2000. "I was really concerned, especially for my family" ("Northern Quebec" 2001). Cooper said the fish that Albert caught and ate were deformed by pollution from a local gold and copper mine. The fishes' eyes were covered with scabs, as many of them also were missing fins.

A report commissioned by the Grand Council of the Cree examined soil and water samples, revealing unhealthy levels of toxic heavy metals, which also were found in hair samples of nearly two dozen Cree people in the same area. The Cree samples for selenium were three times the safe limit, as well as four times higher for lead, and five times higher for aluminum. Hair samples also revealed unusually high levels of arsenic and cyanide. "The levels we have observed particularly of cyanide in fish are killing people and will continue to kill people until something is done," said the report's co-author, Chris Covel ("Northern Quebec" 2001).

In October 2001, the Government of Quebec admitted that it knew the Cree of Ouje-Bougoumou were being poisoned by toxic waste that had fouled their water and poisoned their fish. Canadian government reports documented what the Cree had known for many years: the fish they had been catching and eating were deformed, missing eyes and fins, and poisoned by three nearby mines. Covel's study found high levels of arsenic, cyanide, mercury, and other heavy metals in the water that yielded the deformed fish. Environment Minister Andre Boisclair at first tried to dismiss the findings, but later admitted that the Ouje-Bougoumou Cree were, in fact, being poisoned. Furthermore, he said that the Quebec provincial government had known about the severity of the pollution since at least 1999. "There is a problem," he said. "Their results confirm ours, so there is a problem" ("Quebec Admits" 2001).

"They just dumped it [pollution] into the water," said Chief Sam Bossum, who said that his people could not get the Quebec Environment Ministry to do anything. The Cree then commissioned their own study. United States researcher Chris Covel sampled lake water, mine tailings, lake sediment, fish, and human hair. Analysis indicated high levels of several toxic chemicals. "The levels that we observed, particularly of cyanide in fish, are killing people," he said. As of early 2002, the Government of Quebec had not ordered or funded any actions to remediate the water pollution. Boisclair merely told the Cree "to cut down on their fish intake while he orders more studies" ("Quebec Admits" 2001).

6.12 Colville Tribes Resist Pollution of the Columbia River

Teck Resources Ltd. of British Columbia, which was taken to court by the Colville Confederated Tribes of Washington State for a 100-year record of polluting the Columbia River, was forced to admit in 2012 that runoff from one of the company's smelters had been leaking into the river since 1896. "[It] discharged solid effluents, or slag, and liquid effluent into the Columbia River that came to rest in Washington State, and from that material, hazardous materials [under U.S. environmental laws] were released into the environment," said Dave Godlewski, vice president of environment and public affairs for Teck American. "That's what we've agreed to. We've not talked about the amount of the release. We've not talked about the impacts of those releases. We've just agreed that there has been a release in the U.S." ("Colville Tribes" 2012).

Teck was first sued by the Colvilles in 2004, about 400 tons a day of pollutants, including 250,000 tons of zinc and lead as well as 132,000 tons of other hazardous substances, more than 200 tons of mercury, cadmium, and arsenic ("Watershed Heroes" 2013). "Water is at the heart of who we are as a people, and while there are still significant issues that remain to be resolved in this litigation we are pleased with the stipulation, and believe it is a positive step for the tribes and our river," Tribal Chairman John Sirois said. "The Columbia River is both a national treasure, and the cultural and spiritual center for the Colville peoples. "We hope that this litigation will move us forward as we seek to clean up and protect the health of the Columbia River for all of the people who enjoy and depend on it. We are committed to do all that we can to improve the watershed both for this generation and for the coming generations," Sirois said ("Colville Tribes" 2012).

The plant, 10 miles north of the U.S. border, was dumping into water that entered U.S. territory. On February 23, 2013, the Sierra Club gave the Colvilles its Watershed Heroes Award at a dinner in Spokane, Washington. "We're very grateful for all the sacrifices you have made," Verner said in introducing John Sirois, chairman of the Colville Tribes. "I'm grateful you relayed the history," Sirois said. "I'm grateful for you honoring all the work of the past councils that really put in the time and effort. We have such a great legal team. There are countless people who played a role in this. It's really a validation of who we are as a people. All along the river, those places are named after our people and where we come from: Okanogan, Chelan, Methow, Eniat, San Poil, Lakes, that is who we are. That is where our people are buried. That is where we're born" ("Colville Tribes" 2012).

Start here

6.13 The Penobscots Endure Organochlorine Contamination

Rebecca Sockbeson, a Penobscot, described "the devastating impact of dioxin in my community" to international negotiators of a treaty meant to ban the most widespread Persistent Organic Pollutants (Sockbeson 1999). She said that her nation of almost 500 people live on an island in the Penobscot River, so close to the poisoning effluent of seven pulp-and-paper mills that they are causing significant health problems. Dioxin was being created as a by-product of the chlorine bleaching process that produces paper that is discharged from all of these mills. Dioxin is being poured daily into an adjacent river. Sockbeson said that her people have survived on fish from this river, but that "noow we are dying from it." She continued,

> Neither dioxin nor cancer is indigenous to the Penobscot people, however they are both pervasive in my tribal community. My people face up to three times the state and national cancer rate, moreover, those that are dying of cancer are dying at younger and younger ages, [during] our reproductive generation. This means that unless you take action to eliminate dioxin and other persistent organic pollutants, there will be no Penobscots living on the island by the end of the next century.

The health and survival of Sockbeson's Penobscot band also is threatened by a choice that many Arctic mothers must make: should they breastfeed their children (imparting superior nutrition) and then pass on to them PCBs and dioxins?

"With this," she concluded,

> I humbly, respectfully and desperately urge you to draft a treaty that insures the existence of the Penobscot and other indigenous peoples who are so disproportionately impacted by dioxin. That the breast and spoon we feed our babies with is not filled with cancer, diabetes, learning disabilities, and attention deficit [disorder]." We ask that you please, at this moment, think of your mothers, daughters, sisters, and grandmothers and find the courage to use this unique world power to take action to eliminate dioxin and POPs, creating a promise that I may look upon my great-grandchildren on our common mother, the Earth, as my ancestors look upon me now in this room. *Woliwon*—thank you.

6.14 Oregon Tribes Fight Pollution in and near Portland

Native peoples in Oregon have been resisting "a toxic stew that includes arsenic, cyanide, heavy metals, petroleum, pesticides, dioxins, polychlorinated biphenyls and polycyclic aromatic hydrocarbons. Spanning 12 miles of the Willamette River from the Columbia Slough to the Steel Bridge, the site runs through the heart of the historic port city". The area, at the confluence of the Columbia and Willamette rivers was first listed as an EPA Superfund site in 2000.

Native peoples have sought to protect catfish and bass, as well as marine animals and fish-eating birds that live in or migrate over Portland's harbor. Several species of salmon that are protected by treaties also are at risk. The health of fish passing through Portland's industrial area has become a major issue for Native peoples in the same area. Kelley Point Park also serves Native peoples as a ceremonial site and a launch site for canoe journeys that involve hundreds of people several times a year, as they gather berries, shells, and other traditional foods.

Contributors to the pollution include at least 80 companies and landowners, some of which have gone defunct. In 2013, The EPA said that "cleanup options ranged from $69 million to $2 billion and included a mix of dredging, capping, in-place or after-dredging treatments and use of innovative technologies". The remaining companies (as the Lower Willamette Group) have recommended cleanup options that are the least expensive, but the EPA has found that this group underestimates the risks to human health.

6.15 Coeur d'Alene Demand Cleanup of Mining Waste in Idaho

In October 2001, the EPA proposed a $459 million dollar cleanup of accumulated toxic mining waste in Idaho's Coeur d'Alene River basin. This 1500-square-mile area was devastated in a flood during 1999, after one million pounds of lead flowed into Lake Coeur d'Alene. One million pounds of zinc had previously flowed into the same lake.

The Coeur d'Alene tribe was directly affected by this environmental disaster. The cleanup is expected to require 20 to 30 years at a cost of up to $4 billion. Since 1983, a cleanup has been ongoing near Kellogg, Idaho, which was declared a Superfund site by the EPA. By 2001, more than $200 million had been spent to partially clean this site, most of it to cap mine tailings and replace contaminated soil in local residents' yards ("Hotspots: Idaho" 2001, 3).

The EPA and the state of Idaho seemed to be in no hurry to clean up the site. They consumed more than 30 years to finalize a plan worth $635 million to clean up the upper Coeur d'Alene River basin waste sites, as they grew to be one of the largest Superfund sites in the continental United States. The Bunker Hill Mining and Metallurgical Complex Superfund site, as it is officially known, was first designated in 1983. Mining that had begun during the 1880s leaked more than 100 million tons of waste into the river system. An EPA report said that "until as late as 1968, tailings were deposited directly into the river. Over time, these wastes have spread throughout more than 160 miles of the Coeur d'Alene and Spokane Rivers, lakes, and floodplains" ("Northwest Tribes and EPA" 2012). "This decision sets forth an ambitious, yet thoughtful and methodical approach to reducing risks from metals, making the Coeur d'Alene Basin an even safer, healthier place to live, work and play," said Dan Opalski, director of the EPA's Seattle Superfund office ("Northwest Tribes and EPA" 2012).

In June 2011, the Hecla Mining Company, by then the largest silver miner in the United States, agreed to pay $263.4 million plus interest to the U.S. government, the state of Idaho, and the Coeur d'Alene tribe to clean the Bunker Hill Mining and Metallurgical Superfund Site, one of the United States' largest silver mines between the late 1880s and the late 1960s. The cleanup involved about 100 million tons of mine wastes. The lawsuit that resulted in the settlement was initially filed in 1991. Eventual cleanup may cost $3 billion and require more than 50 years.

According to the EPA, "About 20 miles of streams are unable to sustain a reproducing fish population, and about 10 miles of tributaries have virtually no aquatic life." Children in the area are exposed to hazardous levels of lead in the drinking water, and "more than 15,000 acres of wildlife habitat contain sediments and soils which are acutely toxic to waterfowl" ("Mining Company," 2011).

"Twenty years ago tribal leaders were convinced that not enough was being done to clean up the Coeur d'Alene Basin following a century of mining activity in the Silver Valley," said Chief J. Allan, chairman of the Coeur d'Alene tribe. "Against all odds, the tribe made an unpopular decision to bring one of the largest superfund lawsuits in our nation's history.... The tribe is hopeful that this settlement marks a new chapter in the stewardship of the land we all hold dear. The tribe stands together with the United States, the state of Idaho and Hecla to restore our natural resources while we continue to provide economic prosperity to the region." Roughly three-quarters of the settlement will pay for cleanup begun by state and federal agencies in the mid-1980s (to be concluded in the future). The rest will be "dedicated to the remediation and restoration of natural resources" ("Mining Company" 2011). That will be almost a century to plan and execute the clean-up of a single Superfund site, plus any additional delays along the way.

6.16 The Ramapough: Suffering Ford Motor's PCBs, Heavy Metals, Freon, Arsenic, and Lead

Before they filed a class-action lawsuit against the Ford Motor Company and the EPA in 2006, the Ramapough lawsuit alleged that members of thdethe tribe had suffered several premature deaths from toxic waste dumped by Ford for 30 years. Ford admitted the dumping, but said that it was legal at the time. With help from a documentary by Jamie Redford (son of Robert Redford) shown on Home Box Office, the dumping of Freon, arsenic, PCBs, heavy metals, lead between 1967 and 1971 was publicized. The toxics were dumped by Ford's largest automobile assembly plant, in Mahwah, New Jersey, "into the veritable backyards of the Ramapough Indian people," wrote Vincent Schilling (2011) for the Indian Country Today Media Network.

The communities' children played in the paint sludge while sitting in the landfill areas, painting themselves with a rainbow of dumped paint colors in protest. They also ate handmade paint pies. Members of the community suffered from immediate health problems, such as rashes and nosebleeds and rashes, until nearly every home in the community had had at least one cancer-related death.

6.17 The Yaquis' Borders Don't Stop Pesticide Contamination

The Yaqui are an indigenous farming people who live and work in and near a valley of the same name near Sonora, Mexico. The town and Native neighborhoods cross the United States-Mexico border. Beginning shortly after World War II ended, due to lack of available water and financing, many of the Yaquis became unable to support their own farms as they had for centuries. The Yaquis were then forced to lease their lands to outsiders, including several corporate farmers, who were heavy users of pesticides, herbicides, and fungicides. Use of all of these chemicals, usually applied by tractor, aerial spraying, and by hand, spread a sheen of contamination over the land, water, and people.

Concurrently, valley farm operations became mechanized, as irrigation and transport systems were built. The result was a "Green Revolution," during which farming became big business. Yaqui families from the nearby mountain foothills moved into the valley for employment, while some valley residents moved into the foothills to maintain family-scale farms.

Farmers in the valley reported that two crops a year usually were planted, with pesticides applied as many as 45 times per crop during the growing season. Toxic chemical compounds included several organophosphate and organochlorine mixtures, as well as pyrethroids. Thirty-three different compounds were used to control cotton pests alone between 1959 and 1990. This list also included occasional dousings of DDT, dieldrin, endosulfan, heptachlor, and parathion-methyl endrin. By 1986, 163 different pesticide formulations were being sold in the southern regions of Sonora. Substances banned in the United States, such as lindane and endrin, were readily available to farmers living in the Mexican parts of the valley.

In the valley, pesticide use widespread and haphazard, and continued throughout the year, with little or no inspection or control by governmental agencies. Contamination of people has been documented, with women's breast milk concentrations of lindane, hexachloride, aldrin, heptachlor, benzene, and endrin all above limits of the

Food and Agricultural Organization of the United Nations after one month of a mother's lactation. During 1990, high levels of several pesticides were found in the cord blood of newborns and in breast milk of valley residents. In this area, children are breast-fed and then weaned onto household foods, allowing measurement of toxic chemicals in both mothers' milk and a child's bloodstream.

In addition, household insect sprays were usually applied each day throughout the year in lowland homes. In contrast, the foothill residents maintained traditional intercropping for pest control in gardens, avoiding the sprays. They usually controlled insects in their homes with flyswatters. Most of the hill-dwelling people were exposed to pesticides only when the government sprayed DDT each spring to control malaria.

Angel Valencia (2000), a spiritual leader of the Yaqui tribe in Sonora, Mexico, in the village of Potam, described the effects of these chemicals among valley residents. Valencia spoke as a representative of the Arizona-based Yoemem Tekia Foundation, an affiliate of the International Indian Treaty Council:

> I have seen with my own eyes the effects of daily contact with these pesticides—it burns their skin, they lose their fingernails, develop rashes and in some cases they have died as a result of exposure to these poisons.... The tragedy of this situation makes me both sad and angry—to think of what has been done to the innocent children who are the future of the Yaqui people. They will not be able to grow and develop, as they deserve to.

During the 1990s, Elizabeth Guillette, an anthropologist and research scientist at the University of Arizona, studied the impacts of pesticide exposure on Yaqui children. The studies of Guillette et al. (1998) confirmed the observations of Valencia, who said that exposure to pesticides had "a serious impact on the health and physical and mental development of the children of our villages" (Valencia 2000). Prior to Guillette's research, researchers at the Technological Institute of Sonora in Obregón, Mexico, had shown that children in Sonora's Yaqui Valley often were *born* with harmful concentrations of many pesticides in their blood and were exposed a second time by their mothers' breast milk.

"I know of no other study that has looked at neurobehavioral impacts—cognition, memory, motor ability—in children exposed to pesticides," said neurotoxicologist David O. Carpenter of the State University of New York at Albany. "The implications here are quite horrendous," he said, because the magnitude of observed changes "is incredible—and may prove irreversible" (Raloff 1998). "Although the children exhibited no obvious symptoms of pesticide poisoning, they're nevertheless being exposed at levels sufficient to cause functional defects," observed pediatrician Philip J. Landrigan of Mount Sinai Medical Center in New York (Raloff 1998).

In Guillette et al.'s study (1998), children in the agrarian region, with heavy were compared to children in the foothills, where pesticide use was minimal. The study selected two groups of four- and five-year-old Yaqui children who resided in the

6.17 The Yaquis' Borders Don't Stop Pesticide Contamination

Yaqui Valley of northwestern Mexico. These children shared similar diets, water-mineral contents, cultural patterns, social behaviors, and genetic backgrounds. The major difference was the level of their exposure to pesticides. Guillette adapted a series of motor and cognitive tests into simple games the children could play, including hopping, ball catching, and picture drawing.

The study was constructed in this manner to minimize variables that affect child growth and development. The population had to meet the requirements of similar genetic origin, related cultural and social values and behaviors, and living conditions, all necessary for comparable study and reference groups.

Guillette had assumed that any differences between the two groups would be small and subtle. Instead, she recalled, "I was shocked. I couldn't believe what was happening (Luoma commented:

> The lowland children had much greater difficulty catching a ball or dropping a raisin into a bottle cap—both tests of hand-eye coordination. They showed less physical stamina, too. But the most striking difference came when they were asked to draw pictures of a person.... Most of the pictures from the foothill children looked like recognizable versions of a person. The pictures from most of the lowland children, on the other hand, were merely random lines, the kind of unintelligible scribbles a toddler might compose.... It appeared likely they had suffered some kind of brain damage.

During a follow-up in 1998, two years after her initial visit, Guillette found that both groups (then in primary school) had improved their drawing abilities. While the lowland children's drawings looked more like people, the foothill childrens' drawings were much more detailed. The lowland youths were still encountering motor problems, most often with balance. "Some of these changes might seem minute, but at the very least we're seeing reduced potential," Guillette said. "And I can't help wondering how much these kinds of chemicals are affecting us all" (Luoma 1999).

No differences were found in physical growth patterns of the two groups of children. Functionally, however, Guillette et al. (1998) wrote, "The exposed children demonstrated decreases in stamina, gross and fine eye-hand coordination, 30-minute memory, and the ability to draw a person." Guillette gave children red balloons for successful completion of tasks. "Well over half of the lesser-exposed children could remember the color in the object, and all remembered they were getting a balloon. Close to 18% of the exposed children could not remember anything," and only half could remember they were getting a balloon. "It was quite a contrast," she said (Mann 2000, C-9).

Guillette said that exposed Yaqui children sometimes walked past somebody and punch them without provocation. Otherwise, they usually sat in groups, mainly motionless, and did little or nothing else. Foothill children were almost always busy, usually in group play. "I'd throw the ball to a group of kids. In the valley, one child would get the ball and just play with it himself," Guillette said (Mann

2000, C-9). The foothills children played with a ball as a group. Mothers from the valley also had more problems becoming pregnant and, once pregnancy finally came, they had higher than average rates of premature births, miscarriages, stillbirths, and neonatal deaths.

While it was not possible to determine Yaquis' exposure to individual pesticides, Guillette said that "we know for sure there has been D.D.T. exposure." The Mexican government "does not know what's being used. The farmer does not give out the information. Pesticides are tied to bank loans, and the banks won't reveal what is being used with certain crops. I just assume everything. The other problem is they get a little of this and a little of that and mix it up. It is very important to remember that the situation is no different agriculturally than what you find in California, the Midwest or the East Coast in the U.S." (Mann 2000, C-9).

"Many of these contaminants have similar reactions in the body," Guillette said. "Many disrupt the endocrine system, which regulates body functions, and that's the main reason I looked at subtle changes. The shift may seem slight, but when they occur within a total society, they can have major implications. To me, the approach should not be treatment of the disease or trying to teach compensation for the deficit but to look at the basic problem of contamination" (Mann 2000, C-9).

"Valley children appeared less creative in their play. They roamed the area aimlessly or swam in irrigation canals with minimal group interaction. Some valley children were observed hitting their siblings when they passed by, and they became easily upset or angry with a minor corrective comment by a parent. These aggressive behaviors were not noted in the foothills.... Some valley mothers stressed their own frustration in trying to teach their child how to draw" (Guillette et al. 1998).

Concluding her studies, Guillette raised a question that summarized concerns of parents in the lowlands of their pesticide-ridden valley: "Environmental change has placed the children of the agricultural area of the Yaqui valley at a disadvantage for participating in normal childhood activities. Will they remain at risk for functioning as healthy adults?" (Guillette et al. 1998).

6.18 The Huicholes Live with Pesticides Around the Clock

Exposure to toxic pesticides is one of the greatest risks faced by indigenous (Native) migrant workers in Mexico, where tobacco growers and other agricultural companies use many of the same poisons. Many of the workers are not allowed safety equipment. They also have no access to showers or facilities to wash their clothes after both bodies and clothes have been all but drenched with pesticides. In addition, many of the workers live day and night in the same pesticide-laced fields that they tend and harvest, exposing them to contamination around the clock.

During 1993, for example, an estimated 170,000 fieldworkers arrived in the valleys of Sinaloa during the planting season. Among these workers, roughly 5000 were found to be suffering from toxic contamination because of handling of or prolonged exposure to pesticides that were being used in cultivation (Diaz-Romo and Salinas-Alvarez n.d.).

Of the 35,000 agricultural laborers who worked in the San Quintin Valley of Baja California during 1996, 70% were indigenous. The majority of indigenous migrant workers who worked in the agro-industrial fields of northern Mexico were Zapotecs, Mixtecos, and Triquis from Oaxaca; Tlapanecos, Nahuas, and Mixtecos, from Guerrero; and Purhepechas from Michoacan. According to Estela Guzmán Ayala, women (34%) and children under 12 years of age (32%) constitute two-thirds of the indigenous labor force in the agricultural regions in northern Mexico (Diaz-Romo and Salinas-Alvarez n.d.).

Ruth Franco, a doctor of work-related health and coordinator of the Program for Day Laborers in Sinaloa, estimated that 25% of the roughly 200,000 workers were children between the ages of 5 and 14 during the 1995–1996 cycle in the Sinaloa valley. In the fields where these children and their families work, observers asserted that "thousands of used containers and toxic residues that are generated by the annual use of upwards of 8 million tons of pesticides are criminally disposed of in *ad hoc* trash bins, channels, drains, incinerators, and recycled to storing drinking water" (Diaz-Romo and Salinas-Alvarez n.d.).

The migrants into the Valley included between 15,000 and 20,000 Huicholes who inhabit the mountains of the Sierra Madre Occidental. About 40% of all Huichole families leave their communities during the dry season to find work in the tobacco fields of the Nayarit coast. During the rainy season, the Huicholes usually cultivate a combination of squash, amaranth, corn, chilis, and beans. The Mexican government has made the usual cycle difficult by promoting monocultural planting, distributing hybrid seeds of corn that require the use of pesticides and artificial fertilizers, and replacing the mixed seeds that traditionally were used by the Huicholes and other indigenous agricultural peoples. Use of industrial-scale monocultural agriculture breaks down the indigenous traditions of cooperation among the indigenous peoples, as, at the same time, increasing malnutrition and alcoholism among the workers and their families. Introduction of herbicides such as Paraquat and 2,4-D gradually destroys communal work, and at the same time placing the health of cultivators and their families in danger.

A report on the travails of the Huicholes said, "To arrive at the tobacco fields the Huicholes make a journey from the sierras [mountains] under subhuman conditions, arriving hungry and thirsty. The 'valuable and appreciated' human merchandise includes pregnant women, babies incapable of crying, mute from pain, who have recently been born to malnourished mothers or mothers with tuberculosis. Vulnerable elders and even the 'strong' men arrive at these centers in weak condition" (Diaz-Romo and Salinas-Alvarez n.d.).

Favored workers get purified water, while the remainder have been forced to drink water from irrigation ditches that draw from the pesticide-laced Santiago River or local wells that also are contaminated with the wretched chemical cocktail that is used in the tobacco fields. As they toiled in the heat, the workers become drenched with sweat, allowing their bodies to absorb pesticide residues more easily. Nicotine in the tobacco also causes skin irritation and hives, called "green tobacco sickness."

Child laborers are particularly susceptible to the effects of pesticides and green tobacco sickness.

The harvesting families often spend the entire day and night in the fields, living and sleeping in boxes or under blankets and sheets of plastic beneath strings of drying tobacco leaves, exposing themselves to even more toxic chemicals and tobacco residues. Most have no potable water, drainage, or latrines. Occasionally, "the Huicholes use the empty pesticide containers to carry their drinking water, without paying notice to the grave dangers that this represents, since the majority cannot read the instructions on the labels which may be written in English" (Diaz-Romo and Salinas-Alvarez n.d.).

As with Akwesasne and other Native American communities that have been cursed with persistent organic pollutants, work that concentrates on how to get them out them out of this deadly ecosystem is much less common. Is this because medical work is more likely to receive funding? Or is it for want of a knowledge base? Or is it because the sources and spread of such things are a matter of nature's own creation, with a deadly side of human hubris. A number of studies have found increased rates of melanomas, liver cancer, gall bladder cancer, biliary tract cancer, gastrointestinal tract cancer, and brain cancer, and many others that may be linked to breast cancer.

Several medical studies describe the Inuits' geographical location and traditional lifestyle, in which Canadian Inuit children are highly exposed to polychlorinated biphenyls (PCBs) and lead (Pb), environmental contaminants that are thought to affect fetal and child growth. Investigators examined the associations of these exposures with the fetal and postnatal growth of Inuit children"....child blood PCB concentrations were associated with reduced weight, height, and head circumference during childhood. Cord blood lead was related to smaller height and a tendency to a smaller head circumference during childhood. Results suggested that chronic exposure to PCBs during childhood is negatively associated with skeletal growth and weight, while prenatal lead exposure was related to reduced growth during childhood....This study is the first to link prenatal lead exposure to poorer growth in school-age children, no previous study has investigated the relations of pre- and postnatal PCB exposure to the growth of school-age children.

Reading through reams of professional papers on what POPs can do to the body, especially young bodies, the "punch line" is even more discouraging: if it swims, and if it swims in the Arctic or close to an industrial site, it's probably contaminated. If it is part of a traditional diet, and it swims, these foods are dangerous to eat. All of this is part of a trend that is no more than a few decades old, and the Inuit, had no warning about the results of consuming these traditional foods with these deadly toxins and passing them on to children. What is more, once scientists studied these chemicals and their effects, they learned that other chemicals also mimic the symptoms of PCPs and other POPs.

For example:

6.19 Pregnant Inuit Women more Exposed to "the New PCBs" than Other Canadians

"These contaminants (PFAAs), which are on the increase, are an additional environmental injustice, according to a Université Laval researcher said in the *Montreal Gazette,* October 25, 2020. [https://montrealgazette.com/news/local-news/pregnant-inuit-women-more-exposed-the-new-pcbs-than-other-canadians-study?ct=t(RSS_EMAIL_CAMPAIGN)]

"Pregnant women in the Nunavik region in northern Quebec are twice as exposed to certain chemicals produced far from home than a representative sample of other Canadian women of the same age group," according to a study led by a researcher from Université Laval.

"These contaminants, which are on the increase, are an additional environmental injustice,'" said Mélanie Lemire, a professor in the Quebec City university's medical faculty who is affiliated with the Centre Hospitalier Universitaire de Québec.

"Perfluoroalkyl acids (PFAAs) are chemical compounds used in particular in the manufacture of water-repellent or stain-resistant treatments, non-stick coatings, food packaging, paints, cosmetics and cleaning products....during fetal exposure, these compounds are associated with disturbances of hormonal, renal, cardio-metabolic and immune functions. In my eyes, as an environmental health researcher, these are the new PCBs," Lemire said, referring to polychlorinated biphenyls. [emphasis added]. The import, manufacture, and sale of PCBs have been banned in Canada since 1977 because of the threat they represent to the environment and human health.

Lemire and her colleagues measured changes in the concentration of PFAAs in the blood of 279 pregnant Inuit women between 2004 and 2017. While they observed a decrease in the concentration of regulated PFAAs, they also measured increases of sometimes more than 20% in the concentrations of long-chain PFAAs. These possibly come from the degradation of similar compounds, fluorotelomer alcohols (FTOHs).

"Long-chain PFAAs are more recent molecules, which appeared around 2011 or 2012, and that we also started to measure in Nunavik," Lemire said. "This is where we see an upward trend, while those that are regulated would be downward. Their analyses showed that in 2016-2017, the exposure of pregnant Inuit women to all PFAAs was twice as high as that measured in a comparable sample of other Canadian women." Cross-checking with diet has shown an association between the blood concentration of PFAAs and the consumption of traditional foods. "Traditional foods are at the heart of Inuit culture and the last thing we want is for people to move away from their culture and their food traditions because of contaminants," Lemire said. "The thread is very thin, but they have a right to know. These are the results that belong to them."

Regulations surrounding these compounds are so nebulous that even Lemire is unclear about what they mean. Older PFAAs are regulated internationally. The newer long-chain PFAAs are regulated in North America, but not globally. And since contaminants know no borders, they can accumulate in the Arctic from other parts of the world, such as Asia. Some PFAAs are also found in consumer goods imported into the country.

Not very biodegradable, PFAAs persist in the environment and can be transported over long distances by air or ocean. This is how they find their way into the Arctic food chain and accumulate in the tissues of living organisms as one progresses to the top of the food pyramid. High concentrations of PFAAs are measured in several Arctic wildlife species, despite the absence of significant local sources of emissions. The study found an association between the consumption of traditional seafood and increased exposure to PFAAs. However, Canadians should not be fooled by thinking this problem only affects isolated Inuit populations in the Arctic, the researchers warned. "These are contaminants that are in our consumer goods. This is a question that concerns us all. These contaminants deserve to be studied in different compartments of our population." *The Montreal Gazette's* story was based on a paper published in the scientific journal *Environmental International.* 145: December 20, 106,169, by Élyse Caron-Beaudoinab, Pierre Ayotte Caty Blanchett, Gina Muckle, Ellen Avardg, Sylvie Ricard Mélani Lemire. https://www.sciencedirect.com/science/article/pii/S0160412020321243.Last accessed April 22, 2022.

For the hard-science-inclined, here is the punchline, which indicates that these "new," (that is to say, previously undetected) POPs. Perfluoroalkyl acids (PFAAs) are persistent and ubiquitous environmental contaminants that potentially disrupt endocrine system functions. While some PFAAs perfluoro octane sulfonate (PFOS), perfluorooctanoic acid (PFOA) are regulated, currently used fluorotelomer alcohols (FTOHs) can be transported to the Arctic and are degraded in a number of PFAAs which biomagnify in Arctic wildlife e.g. perfluorononanoic acid (PFNA), perfluorodecanoic acid (PFDA), perfluoroundecanoic acid (PFUdA) *Translation:* Even if it doesn't swim, it still may kill you. Like most POPs, these chemicals can be difficult to detect, and very, very tough to get rid of. You may even find some of it used as a flame retardant in your couch. From a report compiled by the authors of the above study is referenced as ÉlyseCaron-Beaudoinab, et al.

From 2004 to 2017, 279 pregnant Inuit women were recruited as part of biomonitoring projects in Nunavik. The study's goal was to evaluate: (i) time-trends in plasma/serum PFAAs levels in pregnant Nunavimmiut women between 2004 and 2017; (ii) compare plasma/serum PFAAs levels in Nunavimmiut women in 2016–2017 to those measured in women of childbearing age in the Canadian Health Measure Survey (CHMS); and (iii) evaluate the associations of PFAAs levels with the consumption of country foods and pregnancy and maternal characteristics during pregnancy in the 97 participants recruited in 2016–2017....In multivariate models, PFHxS, PFOS, PFNA, PFDA, and PFUdA levels in 2016–2017 were strongly associated with the omega-3/omega-6 PUFA ratio, indicating a positive association between marine country foods consumption and higher exposure to PFAAs.

Conclusions of this study included: The exposure of pregnant women to long-chain PFAAs (PFNA, PFDA, and PFUdA) increased from 2004 to 2017 in Nunavik. Associations noted between PFAAs levels and the omega-3/omega-6 ratio highlight the importance of implementing additional strict regulations on PFAAs and their precursors to protect the high nutritional quality and cultural importance of country foods in Nunavik [emphasis added].

And then there is another twist: POPs seem to be connected to more health maladies than previously suspected. Here a scientific study was summarized by the Canadian Broadcasting Corp. (CBC). See: Jimmy Thomson. "Studies of Inuit and POPs; Scientists Find Link Between Group of Pollutants and Health Problems in Inuit Have Been Finding More Potential Problems Over the Years. New Studies Have Connected High Levels of Persistent Organic Pollutants to Diabetes and High Cholesterol. Canadian Broadcasting Corp. (CBC News), January 19, 2018.

This study found that: A study from the University of Ottawa found a link between persistent organic pollutants (POPs) and high cholesterol levels among the Inuit. The study followed earlier work that found a similar link between the group of chemicals and some types of diabetes. That group of pollutants includes such well-known chemicals as polychlorinated biphenyls (PCBs) and the pesticide DDT. Many of the POPs were outlawed by the international Stockholm Convention. The only problem here is that the POPs do not quit what they do out of respect for international law.

"As researchers, we're concerned with what types of levels are found, and what are the health implications of those exposures," said Kavita Singh, a PhD student who published the recent paper. In Singh's study, PCBs were associated with high levels of total cholesterol and low-density lipoproteins, known wisely as "bad" cholesterol. It adds to the weight of evidence that the chemicals are connected with poor health conditions. "Especially with PCBs, there's a lot of evidence collected from several different populations that they may be linked with chronic diseases, especially diabetes" (Thomson 2018).

6.20 How PCBs Move

The chemicals are released by industries at lower latitudes, but accumulate in the Arctic on prevailing winds, settling out into "sinks"—places in the atmosphere where chemicals descend. They thus become part of the food chain. Smaller animals ingest them and then are eaten by larger ones, their potency increasing geometrically up as the food chain of predators such as seals, and whales, through the food web, concentrating in predators like whales, seals, polar bears, and humans. "A lot of people actually find it surprising that Inuit are exposed to such high levels of contaminants," says Singh, noting that the pollutants travel to the north through the atmosphere and water. "The Arctic ends up acting as a sink" (Thomson 2018).

The Inuit and their food sources never asked for any of this. The chemicals did not ask for it, either. It is a manufactured curse that no one can see, taste, or smell. Because of lower latitude toxicity both the Inuit and their food sources have infested them with incredibly potent, very long-lasting poisons. This is the same set of biological circumstances that have been poisoning Mohawks at Akwesasne, and other peoples who eat POP-laced food.

Despite the dangers, many Inuit prefer traditional "country food" to imported, lower latitude fare. Inuvialuit elder Roy Goose told the CBC (Canadian Broadcasting Corporation) News that "Our bodies are probably riddled with POPs and all kinds of other pollutants in the water column" but "We were meant to eat seal meat and Arctic char and whales from the ocean and waterfowl from the air. That's who we are." Goose says he does not know how he can limit his exposure to pollutants found in most of the foods native to his culture. "Do we become farmers?" he asked. "I think not." Whatever one may believe, the PCBs do not ask a person's cultural preferences before infecting them, and their offspring not yet born, who are another step up the food chain. Medical doctors and scientists can observe it and write papers, but they cannot cure it.

Here are some of the same ideas in more scientific language.

The influence of persistent organic pollutants in the traditional Inuit diet on markers of inflammation

- L. K. Schæbel
- E. C. Bonefeld-Jørgensen
- H. Vestergaard
- S. Andersen
- Published: May 19, 2017
- https://doi.org/10.1371/journal.pone.0177781

Schæbel LK, Bonefeld-Jørgensen EC, Vestergaard H, Andersen S (2017). The influence of persistent organic pollutants in the traditional Inuit diet on markers of inflammation. PLoS ONE 12(5): e0177781. https://doi.org/10.1371/journal.pone.0177781

Other research has indicated that higher POP levels in many Inuit, most notably the elderly, disruption of the immune system, and cardiovascular diseases that are frequent in Greenland Inuit who have been living on the marine diet, meanwhile accumulating PCBs and other chemicals in their body fat.

Abstract Concentrations of persistent organic pollutants (POPs) "are high in Inuit living predominately on the traditional marine diet....We aimed to assess the association between exposure to POPs from the marine diet and inflammation, taking into account other factors such as vitamin D. We invited Inuit and non-Inuit living in settlements or the town in rural East Greenland or in the capital city Nuuk. after adjusting for age, BMI [Body Mass Index], vitamin D, alcohol and smoking. POP levels were associated with the intake of the traditional Inuit diet and with markers of inflammation....This supports a pro-inflammatory role of POPs to promote chronic diseases common to populations in Greenland. These data inform guidelines on 'the Arctic dilemma' and encourage follow-up on the ageing Arctic populations" (Schæbel, et al., 2017).

POPs are highly lipophilic and bio-accumulate in fatty tissues of animals and humans. Hence, bioaccumulation occurs *in addition to* magnification in the food chain that cause high concentrations in top predators. Due to long-range atmospheric transport human exposure to POPs is ubiquitous and not restricted to individuals living in industrial areas, large cities, or parts of the world where POPs are still in use. Populations in the Arctic may even experience larger exposure to POPs than populations living in industrialized parts of the world.

The more Inuit are tested for relationships between POPs and different health problems, the more relationships to POPs and other chemicals have been found. In every case, because POPs reside in the fats of marine animals that comprise "country food," the more they become implicated in various aspects of an expanding Inuit health crisis. "The line of adverse effects of POPs include endocrine disruption within the reproductive system, central nervous system with developmental and behavioral disabilities, cancer, metabolic disorders, the immune system and cardiovascular disease. A main background for cardiovascular disease (CVD) is atherosclerosis, and chronic inflammation has been suggested to be involved in development of atherosclerosis and thus CVD. Previously, we found high levels of markers of inflammation with a frequent intake of the traditional Inuit diet. This diet contains POPs that have been shown to act in a pro-inflammatory manner. POPs may thus contribute to the rise in the occurrence of CVD in Greenland.

This led us to investigate the association between exposure to POPs from the marine diet and inflammation, among Inuit and non-Inuit living in Greenland, taking into account other factors including vitamin D (Schæbel, et al., 2017).

As at Akwesasne, studies from Inuit country (also widely contaminated by PCBs et al.), include many medically-oriented journal articles which describe the malevolent nature of POPs. Very little reporting seems to exist on how to rid the environment of these long-lived toxins. Since they bio-accumulate up the food chain, the most popular argument made by non-Inuit seems to be: quit eating country food, which is to say just about anything that swims. Such a ban presents problems, such as the cost and lack of quality imported food and the preference by most Inuit for culturally relevant country food.

Some nutritional indicators are contradictory. However, while much of Inuit traditional food is high in PCBs and other chemicals from the South, they are also contain beneficial Omega 3-fatty acids. However, the evidence is getting stronger that PCBs and several other long-lasting, bioaccumulating chemicals, some of which were unknown only a few years ago, affect Inuit people, especially the young, with negative health consequences.

More context has been provided by another study:

Verner, Marc André, Pierrich Plusquellec, Justine Laura Desjardins, Chloé Cartier, Sami Haddad, et al. Prenatal and early-life polychlorinated biphenyl (PCB) levels and behavior in Inuit preschoolers.. *Environment International*, Elsevier, 2015, 78, pp. 90–94. ⟨10.1016/j.envint.2015.02.004⟩. ⟨hal-01134338⟩

A team of Canadian researchers, mostly from Quebec, in 2015, found that whereas it is well-established that prenatal exposure to polychlorinated biphenyls (PCBs) can disrupt children's behavior, early postnatal exposure has received relatively little attention in environmental epidemiology. Our study adds to the growing evidence of postnatal windows of development during which children are more susceptible to neurotoxicants like PCBs. (Verner, et al., 2015).

Nature is neutral and non-judgmental, and just as often, painful and lethal.

6.21 The Scope of Knowledge About Effects of Contamination Grows

As time has passed, the number of chemicals tested for effects on Inuit people has widened, and the amount of risk known to scientists has swollen. The amount of contamination in the Arctic has grown from polychlorinated biphenyls (PCBs) to other substances as well. The risk has existed for an unknown length of time, but scientists' tests are catching up with reality. The Arctic's role as a "sink" for toxics is becoming better known.

For example, studies have linked POPs and DDT with Inuits' relatively high levels of persistent organic pollutants to diabetes and high LDL ("bad") cholesterol (Thomson 2018). The University of Ottawa researchers also have been looking at heavy metals such as arsenic and methyl mercury, and how they may affect cardiovascular disease.

While many of these chemicals have been banned or restricted under the international Stockholm Convention, they have persisted in the environment and in the bodies of humans and other animals, bioaccumulating up their food chains. "As researchers we are concerned with what types of levels are found, and what are the health implications of those exposures," said Kavita Singh, a PhD student who published a recent paper that draws on data from the 2007–2008 Inuit Health Survey. Singh's research indicates that "PCBs were associated with high levels of total cholesterol and low-density lipoproteins, the so-called 'bad' cholesterol. It adds to the weight of evidence that the chemicals are connected to poor health outcomes" (Thomson 2018).

Singh also said data indicates that "especially with PCBs, there's a lot of evidence collecting from several different populations that they may be linked with chronic diseases, especially diabetes" (Thomson 2018).

"A lot of people actually find it surprising that Inuit are exposed to such high levels of contaminants," said Singh, noting that the pollutants travel to the north through the atmosphere and water. "The Arctic ends up acting as a sink" Thomson 2018).

Many Arctic peoples are aware that their major food sources have become contaminated with chemical pollution that can be fatal in several ways "Despite the higher levels of pollution in country foods such as seal and whale," Inuvialuit elder Roy Goose told Jimmy Thomson of the Canadian Broadcasting Corp. that he has little choice but to eat what nature provides.

6.22 "Do we Become Farmers?"

"Our bodies are probably riddled with POPs and all kinds of other pollutants in the water column," he said. But, "we were meant to eat seal meat and Arctic char and whales from the ocean and waterfowl from the air. That's who we are." (Thomson

2018). Without them, "do we become farmers?" Even if they tried growing their own food on land, the rocky soil and cold climate would make agriculture impossible.

At the same time that PCBs and other chemicals have been implicated as aggravators of several maladies, another chemical compound, Perfluoroalkyl acids (PFAAs) also occurs in Inuit women at least twice the rate of women south of the Arctic, posing increasing risk and becoming, "an additional environmental injustice," according to a study at Université Laval. Some researchers call them "the new PCBs" (Legault 2020).

Legault quoted Mélanie Lemire, a professor in the Quebec City university's medical faculty who is affiliated with the Centre Hospitalier Universitaire de Québec "These contaminants, which are on the increase, are an additional environmental injustice" (Legault 2020).

The PFAAs have been used notably "in the manufacture of water-repellent or stain-resistant treatments, non-stick coatings, food packaging, paints, cosmetics, and cleaning products. During fetal exposure, these compounds are associated with disturbances of hormonal, renal, cardio-metabolic, and immune functions.

"In my eyes, as an environmental health researcher, these are the new PCBs" (Legault 2020). Import, manufacture, and sale of PCBs were banned in Canada and many other countries after 1977 because of their toxic effects to the environment and human health, especially in the Arctic.

Attesting to their persistence, POPs et al. are still contaminating the bodies of human beings and animals several generations after initial exposure, similar to the herbicide Agent Orange.

The analyses of Lemire and colleagues indicated "that in 2016-2017, the exposure of pregnant Inuit women to all PFAAs was twice as high as that measured in a comparable sample of other Canadian women". Cross-checking with diet has shown an association between the blood concentration of PFAAs and the consumption of traditional foods. "Traditional foods are at the heart of Inuit culture and the last thing we want is for people to move away from their culture and their food traditions because of contaminants," Lemire said. "The thread is very thin, but they have a right to know. These are the results that belong to them" (Legault 2020).

PFAAs are not biodegradable and are easily transportable, covering long distances by wind and ocean currents. Among other places, they "find their way into the Arctic food chain and accumulate in the tissues of living organisms as one progresses to the top of the food pyramid. High concentrations of PFAAs have been measured in several Arctic wildlife species, despite the absence of significant local sources of emissions" (Legault 2020).

This study found an association between the consumption of traditional seafood and increased exposure to PFAAs. However, Canadians should not be fooled by thinking this problem only affects isolated Inuit populations in the Arctic, the researcher warned (Lagault, 2020). "These are contaminants that are in our consumer goods. This is a question that concerns us all. These contaminants deserve to be studied in different compartments of our population (Lagault, 2020). The study's findings were published by the scientific journal *Environmental International* (Caron-Beaudoin et al. 2020).

For example, according to Jane George of *Nunatsiaq News,* an Arctic newspaper, the "Eating the meat of a beluga whale, which is high in PCBs, in turn, are passed on to Inuit who eat the maktaaq (a delicacy of whale skin and blubber). The same is true for the meat and fat of ringed seals, narwhal and polar bears."

"Once again the health of Inuit appears to be under threat from something that people in the North can't see and didn't cause, but which is a menace that's likely to intensify as the Arctic warms," said George.

6.23 Bad News on Whales near the Faroe Islands

A study from the Faroe Islands also found exposure to PCBs is associated with deficient immune function in children, reducing the success of immunizations. Warnings about chemical contamination of large sea creatures that comprise a major part of Inuit diet have come from other places, such as the Faroe Islands. Pál Weihe, a doctor of public health there, advised people to stop eating pilot whale meat and blubber. Skeptical Faroe islanders asserted that his professional practice had been hijacked by animal-rights activists.

That was 1998, when Weihe advised Faroe islanders to limit pilot whale consumption to once a week. Ten years after that, in 2008, having read up on PCB dangers, Weihe expanded his advisory ban to all pilot whale consumption, all of the time. "That's because his new research results showed the heavy load of industrial pollutants in pilot whales can lead to even more serious short and long-term health impacts," said Weihe (George 2011). He also found that eating pilot whale meat and blubber made immunizations for tetanus and diphtheria less effective, raising the risk of serious illnesses.

That was the science that landed Weihe in cultural hot water because whales are not just dinner there. In the Faroes, whaling is said to be the foundation of the national identity. Even so, Weihe said, It's "a reasonable assumption that the findings in the Faroe Islands are representative of the kinds of adverse effects encountered in other communities" (George 2011).

Weigh said he "had a professional obligation to say what we had found and tell people" (George 2011). "Is it politically correct to suppress information?" asked Weihe, who admits his findings made the Faroese government "very nervous." Weihe conceded that a hunter may pursue a whale, just don't eat it. He also said that "Contaminants in pilot whales also appear to increase the risk of Parkinson's disease, hypertension and hardening of the arteries in adults who eat pilot whale regularly". Marine food pollutants also contribute to the development of type 2 diabetes among elderly Faroese residents, Weihe said. The Faroese, whose use of whale meat and blubber as a staple food and a cultural icon for hundreds of years before PCBs (et al.) were developed, had a right to feel stepped on by the industrialized outside world.

6.24 Mercury Also Damages Childrens' IQ

The register of harm that can damage the human body and mind by eating tainted whale meat and blubber does not stop at the Faroe Islands. In Rankin Inlet, belugas are an indigenous delicacy, and the problem (or at least the one that has been tested) is lower IQs in children who have received mercury from mothers' breast milk. "Inuit kids with the highest exposures to mercury in the womb are four times more likely than less-exposed Inuit kids to have low IQs and require remedial education, according to new findings by a team of researchers in Canada and the United States. The children scored on average almost five points lower on IQ tests," wrote Lindsey Konkel in the *National Geographic* (2022). ""This study adds to a wealth of evidence that mercury from seafood can damage brain development in children," said *Philippe Grandjean*, a Harvard University neuroscientist who co-authored landmark research on the *effects of mercury* on children in the North Atlantic's Faroe Islands."

Furthermore, in the world as a whole, mercury levels in oceans are slowly rising, due mainly to coal burning electricity generation in eastern and southern Asia, largely because of emissions from coal-fired power plants in Eastern and Southern Asia. Previous research also has suggested that mercury and PCBs can worsen effects on *attention disorders*, *motor skills*, *heart rates*, and *respiratory* and *ear* infections. Nearly eight of every ten women of childbearing age in some Nunavik villages had blood-mercury levels above Canada's health guidelines. And children with the highest levels were four times as likely to have an IQ below 80, which is generally considered the clinical cutoff for learning disability (Konkel 2022).

All of these harms come from an industrial world that seems to be using its highest-IQ scientists to invent artificial poisons that are contaminating the entire Earth—and, one should add, condemning very large numbers of indigenous peoples around the world to substantially damaged lives for generations to come.

References

Amundsen, Roald. 1908. *The North West Passage*. London: E. P. Dutton & Co.

Blenkowski, Brian. 2013. "Contaminated Tribe: Hormone-Blocking Chemicals Found in First Nation Families." Environmental Health News, November 11. http://www.environmentalhealthnews.org/ehs/news/2013/contaminated-tribe. Last accessed January 22, 2013.

Boucher, O., G. Muckle, J. L. Jacobson, R. C. Carter, M. Kaplan-Estrin, P. Ayotte, É. Dewailly, and S. W. Jacobson. 2014. "Domain-Specific Effects of Prenatal Exposure to PCBs, Mercury, and Lead on Infant Cognition: Results from the Environmental Contaminants and Child Development Study in Nunavik." *Environmental Health Perspectives*. https://doi.org/10.1289/ehp.1206323. Last accessed December 22, 2014.

Brooke, James. 2000. "Canada's Bleak North Is Fertile Ground for Suicide." *Canadian Aboriginal News*, December 18. http://www.canadianaboriginal.com/health/health21a.htm. Last accessed February 3, 2001.

Calamai, Peter. 2000. "Chemical Fallout Hurts Inuit Babies." Toronto *Star*, March 22. http://irptc.unep.ch/pops/newlayout/press_items.html. Last accessed March 31, 2009.

Caron-Beaudoin, Élyse, Pierre Ayotte, Caty Blanchette,, Gina Muckle, Ellen Avar, Sylvie Ricard, and Mélanie Lemire. "Perfluoroalkyl Acids in Pregnant Women from Nunavik (Quebec, Canada): Trends in Exposure and Associations with Country Foods Consumption." Environment International 145 (December, 2020).https://www.sciencedirect.com/science/article/pii/S0160412020321243. Last accessed February 20, 2021.

"Colville Tribes Savor Teck Resources Court Admission That It Polluted Columbia River". 2012. Indian Country Today Media Network, October 9. http://indiancountrytodaymedianetwork.com/article/colville-tribes-savor-teck-resources-court-admission-that-it-polluted-columbia-river-138899. Last accessed November 23, 2012.

Cone, Marla. 1996. "Human Immune Systems May Be Pollution Victims." *Los Angeles Times*, May 13, A-1.

Cone, Marla. 2005. *Silent Snow: The Slow Poisoning of the Arctic*. New York: Grove Press.

"Contaminants Have Variety of Effects on Arctic Baby IQs". 2014. Environmental Health News, February 7. http://www.environmentalhealthnews.org/ehs/newscience/2014/Feb/arctic-baby-iqs. Last accessed March 8, 2014.

Dewailly, E., P. Ayotte, S. Bruneau, S. Gingras, M. Belles-Isles, and R. Roy. 2000. "Susceptibility to Infections and Immune Status in Inuit Infants Exposed to Organochlorines." *Environment Health Perspectives* 108: 205–11.

Dewailly, E., S. Bruneau, C. Laliberte, M. Belles-Iles., J. P. Weber, and R. Roy. 1993. "Breast Milk Contamination by PCB and PCDD/Fs in Arctic Quebec: Preliminary Results on the Immune Status of Inuit Infants." *Organohalogen Compounds* 13: 403–6.

Dewailly E., J. J. Ryan, C. Laliberte, S. Bruneau, J. P. Weber, S. Gringras, and G. Carrier. 1994. "Exposure of Remote Maritime Populations to Coplanar PCBs." *Environmental Health Perspectives* 102, suppl. 1: 205–9.

Diaz-Romo, Patricia, and Samuel Salinas-Alvarez. n.d. "Migrant Workers and Pesticides. A Poisoned Culture: The Case of the Indigenous Huicholes Farm Workers." *Abya Yala News: The Journal of the South and Meso-American Rights Center*. http://saiic.nativeweb.org/ayn/huichol.html. Last accessed December 22, 2002.

Dobbyn, Paula. 2000. "Contaminated Game Has Natives Worried." *Anchorage Daily News*, May 2. http://www.adn.com. Last accessed May 24, 2000.

George, Jane. Contaminants in Whale Meat and Blubber Weaken Childhood Immunizations: Public Health Expert. *Nunatsiaq News*. May 3, 2011. https://nunatsiaq.com/stories/article/03556_contaminants_in_whale_meat_and_blubber_weaken_childhood_immunizations/. Last accessed May 20, 2022.

Guillette, Elizabeth A., Maria Mercedes Meza, Maria Guadalupe Aquilar, Alma Delia Soto, and Idalia Enedina Garcia. 1998. "An Anthropological Approach to the Evaluation of Preschool Children Exposed to Pesticides in Mexico." *Environmental Health Perspectives* 106, no. 6 (June). http://www.anarac.com/elizabeth_guillette.htm. Last accessed August 10, 1998.

Hansen, Terri. 2014. "532 Superfund Sites in Indian Country." Indian Country Today Media Network, June 17. http://indiancountrytodaymedianetwork.com/2014/06/17/532-superfund-sites-indian-country-155316. Last accessed June 25, 2014.

"Heavy Metal Levels in Reindeer, Caribou, and Plants of the Seward Peninsula". 2000. Current Research Programs, Reindeer Research Program, University of Alaska Fairbanks, April. http://reindeer.salrm.alaska.edu/research.htm. Last accessed June 5, 2000.

Hill, Miriam. 2001. "Iqaluit's Waste Woes Won't Go Away; City Sets Up Bins Where Residents Can Dump Plastics, Metal." *Nunatsiag News*, July 27, 2001.

"Hotspots: Idaho". 2001. *Drillbits and Tailings* 6, no. 7 (October 31): 3.

Hulen, David. 1994. "Hunt Is On for Pollutant Traces in Bering Sea; Alaska Villagers, Scientists Wonder If Toxic Substances Are Endangering Animals and People Who Eat Them." *Los Angeles Times*, August 15, A-5.

References

"Iqaluit Drunks Keep Cops' Hands Full". 2001. CBC North/News, March 16. http://north.cbc.ca/cgi-bin/templates/view.cgi?/news/2001/03/16/16iqalcohol. Last accessed April 16, 2001.

Johansen, Bruce E. 2000. "Pristine No More: The Arctic, Where Mother's Milk Is Toxic." *The Progressive*, December, 27–29.

Konkel, Lindsey. "How Brain-Damaging Mercury Puts Arctic Kids at Risk." *National Geographic*, March 26, 2022. www.nationalgeographic.com/science/article/150327-inuit-mercury-beluga-iq-canada-nunavik-arctic-faroe-islands. Last accessed May 18, 20221.

Lamb, David Michael. n.d. Transcript, Canadian Broadcasting Corporation News: Toxins in a Fragile Frontier. http://cac.ca/news/indepth/north/. Last accessed November 30, 2002.

Legault, Jean-Benoit. "Pregnant Inuit Women More Exposed to 'the New PCBs' Than Other Canadians: Study." Presse Canadienne. In *Montreal Gazette*, October 25, 2020. Last accessed December 18, 2020. https://montrealgazette.com/news/local-news/pregnant-inuit-women-more-exposed-the-new-pcbs-than-other-canadians-study

Luoma, Jon R. 1999. "System Failure: The Chemical Revolution Has Ushered in a World of Changes. Many of Them, it's Becoming Clear, Are in Our Bodies." *Mother Jones*, July/August. http://www.motherjones.com/mother_jones/JA99/endocrine/html. Last accessed September 2, 1998.

Mann, Judy. 2000. "A Cautionary Tale about Pesticides." *Washington Post*, June 2, C-9.

"Mercury Poisoning Five Decades Later for Grassy Narrows and White Dog First Nation". 2012. Indian Country Today Media Network, June 12. http://indiancountrytodaymedianetwork.com/article/mercury-poisoning-five-decades-later-for-grassy-narrows-and-white-dog-first-nations-117843. Last accessed July 2, 2012.

"Mining Company to Pay Coeur d'Alene, State of Idaho and U.S. Government". 2011. Indian Country Today Media Network, June 16. http://indiancountrytodaymedianetwork.com/article/mining-company-to-pay-coeur-d%E2%80%99alene%2C-state-of-idaho-and-u.s.-government-38602. Last accessed June 29, 2011.

Mofina, Rick. 2000. "Study Pinpoints Dioxin Origins: Cancer-causing Agents in Arctic Aboriginals' Breast Milk Comes from U.S. and Quebec." Montreal *Gazette*, October 4, A-12.

"Northern Quebec Cree Poisoned by Mining Pollutants: Report." Canadian Broadcasting Corporation News Online, October 20. http://cbc.ca/cgi-bin/view?/news/2001/10/19/cree_quebec011019. Last accessed November 19, 2001.

"Northwest Tribes and EPA Agree on Cleanup Plan for Upper Coeur d'Alene River Basin". 2012. Indian Country Today Media Network, September 4. http://indiancountrytodaymedianetwork.com/article/northwest-tribes-and-epa-agree-on-cleanup-plan-for-upper-coeur-d%E2%80%99alene-river-basin-132517. Last accessed September 22, 2012.

PCB Working Group. n.d. "Communities Respond to PCB Contamination." http://www.ipen.org/circumpolar2.html. Last accessed December 30, 2001.

Pugliese, David. 2001. "An Expensive Farewell to Arms: The U.S. Has Abandoned 51 Military Sites in Canada. Many Are Polluted, and Taxpayers Are Paying Most of the $720 Million Cleanup Cost." *Montreal Gazette*, April 28, B-1.

"Quebec Admits Toxic Waste Poisons Cree". 2001. Canadian Broadcasting Corporation News Online, October 26. http://cbc.ca/cgi-bin/view?/news/2001/10/26/cree_tox011026. Last accessed November 16, 2001.

Raloff, J. 1998. "Picturing Pesticides' Impacts on Kids." *Science News* 153, no. 23 (June 6):358.

Schilling, Vincent. 2011. "Documentary Featuring the Fight between Ford and Ramapough Indians Premieres Tonight." Indian Country Today Media Network, July 18. http://indiancountrytodaymedianetwork.com/article/documentary-featuring-the-fight-between-ford-and-ramapough-indians-premieres-tonight-43261. Last accessed August 2, 2011.

Sockbeson, Rebecca. 1999. "Statement by Rebecca Sockbeson for IRATE (Indigenous Resistance Against Tribal Extinction), and I.E.N. (Indigenous Environmental Network)." International POPs Elimination Network, PCB Working Group, September 8. http://www.ipen.org/irate.html. Last accessed October 21, 1999.

Suzuki, David. 2000. "Science Matters: POP Agreement Needed to Eliminate Toxic Chemicals." December 6. http://www.davidsuzuki.org/Dr_David_Suzuki/Article_Archives/weekly120 60002.asp. Last accessed December 28, 2000.

Thomson, Jimmy. "Scientists Find Link Between Group of Pollutants and Health Problems in Inuit; New Studies Have Connected High Levels of Persistent Organic Pollutants to Diabetes and High Cholesterol" CBC [Canadian Broadcasting Corp.] News, January 19, 2018. https://www.cbc.ca/news/canada/north/scientists-find-link-between-group-of-pollutants-and-health-problems-in-inuit-1.4494136. Last accessed May 18, 2022.

Valencia, Angel. 2000. "Statement of Angel Valencia, Yoemem Tekia Foundation, Tucson, Arizona POPs Negotiations, March 23, 2000." Native News: The Mail Archive. March 27. http://www.mail-archive.com/natnews@onelist.com/msg00572.html. April 16, 2000. Last accessed May 25, 2000.

"Watershed Heroes: Colville Confederated Tribes Win Sierra Club Award for Battling British Columbia Smelter". Indian Country Today Media Network. March 6, 2013. http://indiancountrytodaymedianetwork.com/2013/03/06/watershed-heroes-colville-confederated-tribes-win-sierra-club-award-battling-british. Last accessed March 13, 2013.

York, Geoffrey. 2001. "Russian City Ravaging Arctic Land." *Toronto Globe and Mail*, July 25, A-1, A-11.

Chapter 7
Alberta's Moonscape: If This Sounds Apocalyptic, It Is

By now, any astute student of Native American/industrial conflict on the Great Plains of the United States and Canada must understand that a large part of that conflict involves pipelines and what they carry (usually crude oil, natural gas, or some derivation of them), often for export. Native American resistance to this juggernaut nearly always involves the threat of spills and danger to a natural environment that the Native peoples hold sacred and use for day-to-day survival.

The Keystone Pipeline project I (now stalled, at least for a time) has been one the best known of many pipeline challenges across Indian Country. Canceled (or at least substantially delayed) by President Joe Biden's administration after it had been reinstated by Donald J. Trump, after cancelation by Barack Obama, the "XL's" advocates have been waiting for a new Republican president to bring it back to life. It has, during the better part of a generation, become a stalking horse for environmental advocates for its threats to aquifers and other water sources in a dry land, as well as anticipated additions to the world atmosphere's carbon-dioxide load.

Not a day goes by without a challenge to a pipeline somewhere in Indian country. The cast is usually set when Corporation X proposes to move Resource Y through Native Land from Mining Site Z to a refining site at or near a port that very probably will export the resource often across an ocean, to another country. The main problem is usually current or prospective spills of the resource that have or may ruin animals' habitat as well as the lives of Native peoples who live in its proposed path.

7.1 Many Pipelines, Not Much Publicity

Many of these conflicts do not get much publicity outside media with specific Native audiences. For example, witness Enbridge's court battle over a pipeline route through Native lands in Wisconsin: Enbridge wanted to keep oil flowing through about 12 miles of its 645- mile-long oil and natural gas pipeline that runs through Bad River lands.

"Enbridge has been getting personal with the Bad River Ojibwe tribe over the company's Line 5 pipeline route through tribal lands in Wisconsin," read one account in *Indian Country Today* (Pember 2022). "In recent lawsuits, Enbridge has targeted individual tribal members and staff, seeking the court's permission to question them under oath about their "thought process" in opposing renewal of the company's easement through the reservation.

7.2 The Land Is Sacred, but Also a Battlefield

"For Bad River citizens and leaders, however, the issue has always been personal. Bad River or *Mashkiziibii* (Medicine River) has an abiding, irremovable quality for Ojibwe people," the account continued. "Central to their world view and spirituality, and an example of their sustainable connection with traditional foods and ways, Bad River is more than geography. The river and land represent Ojibwe blood memory, according to Aurora Conley, a citizen of the Bad River tribe and a member of the Anishinaabe Environmental Protection Alliance" (Pember 2022).

This battle contains all of the elements common to pipeline disputes across the continent: a company's diehard pursuit of a pipeline route to carry oil through Native land that collides with reservation residents' protection of their rights to clean water and spiritual sustenance on land that has been their home for hundreds of years, or longer.

In this case, the court ruled against the company, but legal battles did not end there. They hardly ever do, leaving the Native residents in a perpetual state of legal limbo. "Ours is a long memory of the fact that we've been here for thousands of years," said Bad River Band Chairman Mike Wiggins. "We have a long vision forward rooted in water resources, in the quality and purity of our water" (Pember 2022).

The pipeline was 71 years old in 2021 and risky: "Line 5 opponents point to Enbridge's history of pipeline leaks and failures. The line has leaked at least 29 times between 1968 and 2017, according to data published by the National Wildlife Federation" (Pember 2022). One of its spills involved 843,000 gallons of tar sands oil.

The Keystone XL Pipeline is not the only conduit for tar sands oil. One major reason that it has been a target of protest is that it crosses an international border, and a high possibility of damaging spills.

"While the risk of a rupture or leak of Line 5 is significant along the entire reservation corridor, the circumstances just east of the location where the pipeline currently passes beneath the Bad River portend a looming disaster," the suit says, according to Pember's account (2022). According to the lawsuit, the Bad River is carving away its banks and soils that conceal and protect the pipeline and will soon expose it to damage.

7.3 A Perpetual Battle

Native peoples are thus engaged in a perpetual battle to protest the industrial economy's encroachment into their lives Every so often, the modern protests grow in size and end up engaging huge world-wide audiences. Such was the occupation of Wounded Knee (1972–1973), which ended as an armed siege in which people on both sides died of gunshot wounds. Closer to our time, a large encampment protested the extension of the Dakota Access pipeline across the United States' northern plains (partly over Native-owned land) in 2016 and 2017. While gunpowder was scarce during this confrontation, it was full of made-for-TV images that raised shivers world-wide, as authorities carrying fire-engine-caliber hoses scattering protesters with nearly freezing high-velocity water on ice-cold Dakota winter days.

The sheer familiarity of environmental damage undergirds Native American peoples' fierce opposition to the Dakota Access Pipeline project. The determined nature of this resistance across North America stems from Native lands' long history as resource colonies which host more than a third of the United States' Superfund acute pollution sites (Johansen 2016).

Environmental provocations afflicting Native American peoples in the United States range from uranium to kitty litter—a range of problems equal to those of any Third World nation—from the toll of uranium mining on Navajo, land, to the devastation wrought by dioxin, PCBs, and other pollutants on the agricultural economy of the Akwesasne Mohawk reservation in northernmost New York State. As with the Akwesasne Mohawks, some of the most serious problems span international borders. The Yaquis, whose homelands span the U.S.–Mexican border, have been afflicted with some of the same chemicals as the Mohawks on the U.S.–Canadian border.

Many reservation residents suffer from cancers and others with illnesses because their lands have been used for several decades as industrial dumps and mine sites. These include acute effects of exposure to dioxins, PCBs, and other persistent organic pollutants, most acutely in the Arctic, where Native consumption of sea life (their traditional diet) has been curtailed and Inuit mothers are sometimes warned not to breast-feed their infants because their milk may be toxic. The Arctic, which looks so pristine to the untutored eye, is also experiencing the world's most rapid rate of climate change, as temperatures rise and ice melts. An Inuit culture based on ice also is melting. and many Inuit hunters have been injured or killed by falling through thin ice.

Canada also has become a major source of indigenous environmental contamination and conflict. The Innu of Labrador have been afflicted with sulfide mining, aluminum smelting, and noise pollution from squadrons of military aircraft. Some of the most intense resource exploitation in Canada takes place in remote locations, such as among the Lubicon Cree of northern Alberta, whose lands were so inaccessible in 1900 that treaty makers completely missed them. Today, roads have opened their lands to massive oil drilling and logging.

7.4 Tar Sands' Devastation in Northern Alberta

Native peoples in Alberta have found some of their homelands devastated by tar (or oil) sand mining to the point that the landscape has been compared to a moonscape. Native peoples in the United States and Canada also have taken a leading role in opposing (and, by 2022, stopping) the Keystone XL Pipeline, which was being proposed (until it was canceled by President Joe Biden) to carry tar sands oil from Alberta to the U.S. Gulf Coast for refining.

During the battle to stop the Keystone XL, Roads carrying equipment to the tar sands fields were blocked, and several arrests took place on Nez Perce land in Idaho and the Lakotas' Pine Ridge reservation in South Dakota. Opposition to tar sands development and the Keystone XL became robust because completion of the pipeline would have allowed tar sands to be further developed, leading to higher greenhouse gas levels in the atmosphere. By 2013, one-third of Alberta's economy was tied in some way to the tar sands (Lizza 2013, 47), including royalties worth as much as $4 billion a year. Environmentalists around the world considered the trade of this crudest of crude oils for money to be a climatic faustian bargain, trading the Earth's future for quick profits, even if its mining disfigures the face of the Earth in ways that are ugly even next to those of modern strip mining of coal, other fuel so other minerals and energy.

7.5 Promotion and Erosion of Support for Tar Sands Oil

Canadian Prime Minister at that time, Stephen Harper, promoted tar sands development avidly, with plans to approve construction and operation for 10,000 miles of pipelines eastward and westward across Canada, as well as the better-known Keystone XL southward and eastward from the tar sands fields in Alberta to the Texas Gulf Coast. By 2014, the Canadian federal Parliament, with Harper's backing, had revoked or annulled 70 environmental laws. In its 2013–2014 budget, the Canadian federal government set aside about $22 million to promote tar sands–based oil outside of Canada (Leslie 2014, A-19). In 2014, Canadian oil producers expected new pipeline capacity to nearly double Canada's production of tar sands oil by 2025 from 3.5 million to 6 million barrels a day, mostly for export (Krauss and Austen 2014, B-4).

Indigenous opponents of both tar sand mining and the Keystone XL Pipeline united with other environmental groups to contend that the amount of fossil-fuel energy required to produce usable energy (such as gasoline) from tar sands exceeded that of conventional oil, contributing to higher levels of greenhouse-gas emissions.

7.6 Tar Sands and Strip Mining

Tar sands have a consistency of gritty peanut butter and can be so thick that they must be thinned (using fossil-fuel energy) to be refined into usable fuel. Most usable tar sands are under ground. Bringing them to the surface often requires a type of strip mining that scars the Earth in ways that cannot be quickly or easily repaired. Opponents of tar sands mining contend that oil sands are a new form of fossil fuel—the last thing the Earth needs when carbon-dioxide levels in the atmosphere have risen to more than 450 parts per million (ppm), more than 50% higher than peak preindustrial levels. With the proportion of carbon dioxide in the atmosphere steadily rising even without adding more by processing and using tar sands, advocates of this new source of energy are also going to be responsible for advancing the dates and potency of climate change, including higher temperatures, rising seas, and oceanic acidity.

On a local level, many people worry about oil spills from pipelines in fragile areas, such as the Nebraska Sand Hills, that could contaminate the Ogallala Aquifer where water is already scarce. This aquifer supplies 78% of public water for human use and 83% of irrigation water in Nebraska, almost a third of the irrigation water used in the United States.

7.7 Tar Sands Mining as the "Fuse to the Biggest Carbon Bomb on the Planet"

Why was the opposition to tar sands and the Keystone XL ("Express Line" to its supporters or "Extra Leaky" to its opponents) so passionate among both Native peoples and many environmentalists? James Hansen, retired director of NASA's Goddard Institute for Space Studies (who was twice arrested at the U.S. White House fence in civil disobedience actions opposing the Keystone XL), has said that full refining and use of tar sands (including export to other countries around the world) would mean "game over" for a habitable Earth. Hansen in 2011 joined 19 other scientists in a letter to President Barack Obama, saying. "We can say categorically that it's not only not in the national interest [of the United States], it's also not in the planet's best interest" ("NASA Scientist" 2011). A "No Tar Sands Caravan" to protest the pipeline traversed the United States from California, arriving in Washington, D.C., on August 29, 2011, after a 3300-mile journey.

According to NASA, the Alberta tar-sands fields, which were first exploited in 1967, are "the world's largest oil-sands deposit, with a capacity to produce 174.5 billion barrels of oil—2.5 million barrels per day for 186 years ("Athabasca Oil Sands" 2011). (The United States as a whole consumes about 15 million to 20 million barrels of oil-derived products per day.) Environmental activist Bill McKibben has called tar sands mining and the "fuse to the biggest carbon bomb on the planet" (Tollefson 2013). "Saying that the tar sands are not necessarily worse than coal is like saying that drinking arsenic is not necessarily worse than drinking cyanide," said geophysicist Raymond Pierrehumbert of the University of Chicago. He said that fully developing the tar sands could by itself, ("even if we suddenly stopped burning coal,") warm the atmosphere an additional 3.6 degrees Fahrenheit by century's end—an amount that climate scientists warn could be catastrophic" (Koch 2014). In other words, we face a future in which oil supplies are not an issue.

The existential issue here is not a need for oil. Hopefully, we will progress to forms of energy that do not ruin the Earth and raise its temperatures to intolerable levels as necessary side-effects. The real survival issue for the Earth, with its plants and animals (including human beings), is how much higher a concentration of carbon dioxide can the atmosphere tolerate before the climate becomes intolerable. The atmosphere's level of CO2 is already as high as during the Pliocene, 2–3 million years ago, when air temperatures were roughly 4–5 degrees C. higher than today. The difference is mainly a matter of thermal inertia, by which actual atmospheric heating occurs 50–200 years after CO2 and other greenhouse gases are pumped into the atmosphere.

7.8 A Signature Environmental Issue

During the debate over it, the Keystone XL pipeline became the signature environmental issue of the Obama presidency. Several large U.S. environmental and Native American groups called Obama's decision on the Keystone XL a "watershed moment." In the meantime, scientists who had stated their support for Obama's position doubled down in similar letters to Trump and Biden:

"Dear President Obama," they wrote, "Many of the organizations we head do not engage in civil disobedience; some do. Regardless, speaking as individuals, we want to let you know that there is not an inch of daylight between our policy position on the Keystone Pipeline and those of the very civil protesters being arrested daily outside the White House. This is a terrible project—many of the country's leading climate scientists have explained why in their letter last month to you. It risks many of our national treasures to leaks and spills. And it reduces incentives to make the transition to job—creating clean fuels," the letter states ("NASA Scientist" 2011).

Signers included the Sierra Club, the Natural Resources Defense Council, Greenpeace, the National Wildlife Federation, Friends of the Earth, the Rainforest Action Network, 350.org, the League of Conservation Voters, Environment America, and the Center for Biological Diversity.

Many of these groups (Greenpeace being most active) also advocate a shutdown of the tar sands fields themselves "and end[ing] the industrialization of a vast area of Indigenous territories, forests and wetlands in northern Alberta.... The tar sands are huge deposits of bitumen, a tar-like substance that is turned into oil through complex and energy-intensive processes that cause widespread environmental damage— polluting the Athabasca River, lacing the air with toxins and turning farmland into wasteland," Greenpeace said. "Large areas of boreal forest would have been clear-cut to make way for development the tar sands, the fastest-growing source of greenhouse gas emissions in Canada" ("NASA Scientist" 2011).

7.9 Tar Sands as Junk Energy

Tar sands are a mixture of clay and sand with bitumen, a thick, low-grade form of petroleum similar to asphalt. According to Thomas Homer-Dixon, who teaches global governance at the Balsillie School of International Affairs, "Tar sands production is one of the world's most environmentally damaging activities. It wrecks vast areas of boreal forest through surface mining and subsurface production. It sucks up huge quantities of water from local rivers, turns it into toxic waste, and dumps the contaminated water into tailing ponds that now cover nearly 70 square miles" (Homer-Dixon 2013).

Also, wrote Homer-Dixon (2013), "bitumen is junk energy. A joule, or unit of energy, invested in extracting and processing bitumen returns only four to six joules in the form of crude oil. In contrast, conventional oil production in North America returns about 15 joules. Because almost all of the input energy in tar sands production comes from fossil fuels, the process generates significantly more carbon dioxide than conventional oil production." According to NASA, oil sand refining produces the equivalent of 86–103 kilograms of carbon dioxide for every barrel of crude oil produced. By comparison, 27–58 kilograms of carbon dioxide are emitted in the conventional production of a barrel of crude oil.

By 2012, the mining and refining of tar sands in Canada consumed as much natural gas as that country used for home heating (Kolbert 2007, 49). The gas is used to produce synthetic oil and by-products, such as gasoline. Tar sands require about 15–40% more energy in manufacture compared to conventional crude oil; oil shales require about twice as much. However, converting tar sands into oil costs about as much as oil, at $30 a barrel. With oil pushing $100 a barrel in 2022, tar sands were becoming very profitable for many fossil-fuel companies (Kolbert 2007, 49, 50).

Ryan Lizza (2013, 42), writing in *The New Yorker*, described the mining of tar sands:

> Oil sand has the texture of soft asphalt; 20% of it lies close to the surface, and the area is effectively strip-mined. The bitumen-rich sand is removed, mixed with water into a slurry, and spun in centrifuges until the oil is separated, leaving behind vast black tailings ponds that are hazardous to wildlife. The mining operations sprawl ruinously for miles. The remaining 80% of the oil sands lie hundreds of feet down beneath a layer of hard rock. Steam is injected deep below ground until the oil naturally separates and is drawn out. The extra energy required to extract the sand makes it a more carbon-intensive fossil fuel—averaging 17% more ... than conventional oil.

7.10 Strip Mining a Moonscape

Berry (2012) of the Indian Country Today Media Network described the environmental damage of tar sands mining:

> If you can imagine the bleak landscape of the moon, you can envision the desolate, 54,000-square-mile tar sands of northern Alberta.... "It's literally a toxic wasteland—bare ground and black ponds and lakes—tailings ponds—with an awful smell," said Warner Nazile ... [an activist] from British Columbia and member of the Wet'suwet'en First Nation. The mining is "despoiling an area roughly the size of England...." University of Alberta scientists "found indications that contamination from the tailings ponds was polluting a huge aquifer [containing underground water] that ultimately flows into the Arctic Ocean," Nazile said. Two aboriginal communities downstream from the oil sands have experienced higher-than-average rates of cancer and other health problems, he added.

The Alberta oil sands in 2012 were being mined across an area of 54,900 square miles, according to a satellite data analysis by NASA. About 80% of the tar sands lie more than 75 meters below ground, where they are extracted by injecting hot water into a well that liquefies the oil for pumping. The NASA analysis found that pollutants from the mines "are comparable to the emissions from a large power plant or a moderate-sized city," including nitrogen dioxide and sulfur dioxide ("Emissions" 2012). Canada has strict environmental laws requiring the restoration of mine sites. By 2011, one tailings pond had been restored, surrounded by a scarred landscape. The large tailings pond from the first mine, opened in 1967, was drained, filled in, and planted with grass, but NASA satellites did not show grass growth as of 2011. Two tons of sand are required, along with water, to produce one barrel of crude oil. Huge trucks remove as much as 720,000 tons of sand per day. The energy required to mine and separate oil from sand adds to its carbon footprint. The mining operation also consumes copious amounts of electric power.

7.11 Tar Sands and Climate Change: "If This Sounds Apocalyptic, It Is"

Before it was twice canceled, the Keystone XL pipeline was designed to carry as much as 800,000 barrels of tar sands crude oil daily from Alberta's strip mines to refineries in Oklahoma and Texas. According to plans, it would have been routed through the Bakken shale formation of North Dakota and Montana, which has been producing new oil from hydraulic fracturing (known as "fracking.") in which high-pressure chemicals mixed with waster are injected deep underground to break apart

7.12 Tar Sands and Oil Spills

the shale rock layer and pump oil out. Tar sands and fracking for oil tap relatively new sources of fossil fuels that may increase in years to come, raising greenhouse gas levels in the atmosphere. James Hansen, author of *Storms of My Grandchildren,* wrote in the *New York Times*:

> Canada's tar sands, deposits of sand saturated with bitumen, contain twice the amount of carbon dioxide emitted by global oil use in our entire history. If we were to fully exploit this new oil source, and continue to burn our conventional oil, gas and coal supplies, concentrations of carbon dioxide in the atmosphere eventually would reach levels higher than in the Pliocene era, more than 2.5 million years ago, when sea level was at least 50 feet higher than...now. That level of heat-trapping gases would assure that the disintegration [melting] of the ice sheets would accelerate out of control. Sea levels would rise and destroy coastal cities. *Global temperatures would become intolerable. Twenty to 50% of the planet's species would be driven to extinction. Civilization would be at risk.* ... *If this sounds apocalyptic, it is.* (Hansen 2012, emphasis added.)

The concentration of carbon dioxide in the atmosphere has risen from 280 ppm to more than 420 ppm over the past 150 years, as of 2022. The tar sands contain enough carbon—240 gigatons—to add 120 ppm to that level. Tar shale, similar to tar sands found mainly in the United States, contains at least an additional 300 gigatons of carbon. If we turn to these dirtiest of fuels, instead of finding ways to phase out our addiction to fossil fuels, there is no hope of keeping carbon concentrations below 500 ppm—a level that would, as Earth's history shows, leave our children a climate system that is out of their control, searingly hot, and getting hotter.

In an attempt to address critics of oil sands' carbon emissions, Alberta's government has invested in a $1.35 billion demonstration project to capture and bury as much as 1.2 million tons of carbon dioxide a year from an oil sands facility that manufactures oil from bitumen. Alberta also has required reductions in carbon intensity (emissions per unit of production) by 12%, or payment of a $14-per-pound carbon tax. Between 2007 and 2012, that tax raised $378 million for clean energy investments, of which about $80 million was spent on projects related to oil sands (Tollefson 2013).

7.12 Tar Sands and Oil Spills

Opponents of the Keystone XL point out the less-than-sterling record of existing pipelines, especially recent leaks in several communities and rural areas that were bearing high cleanup costs even without added pipeline capacity, such as an Enbridge Energy spill near Marshall, Michigan, in 2010, in which more than

840,000 gallons of tar sands crude oil was spilled. In March 2012, an Exxon pipeline sprang leaks in Mayflower, Arkansas, a town of 2200 people near Little Rock, Arkansas, resulting in the evacuation of about two dozen homes. This pipeline also was carrying about 210,000 gallons of tar sands crude oil that spilled.

Both companies spent tens of millions of dollars trying to recover the heavy crude, similar to what the Keystone XL would have carried. River and floodplain ecosystems have had to be restored, and neighborhoods refurbished. Legal battles were waged, and residents' lives changed for a very long time.

"All oil spills are pretty ugly and not easy to clean up," said Stephen K. Hamilton, a professor of aquatic ecology at Michigan State University, who advised the Environmental Protection Agency (EPA) and the state on the cleanup in Marshall. "But this kind of [tar-sands] oil is even harder to clean up because of its tendency to stick to surfaces and become submerged" (Frosch 2013). A 40-mile segment of the river that traverses Marshall, one site of the Enbridge pipeline, was closed for 2 years, and several waterfront homes were abandoned. In the meantime, Enbridge underestimated the amount of oil that had been spilled by a factor of about 100. That is: for each gallon of thick, sticky oil that industry "experts" *thought* would spill, about 100 gallons actually *did* enter the ecosystem. With a record such as that, industry "experts" could very easily be mistaken for oil-industry public-relations shills.

The EPA ordered the company to dredge the river. As a show of its contrition, Enbridge also bought out 154 homes in the area most intensely affected by the spill and spent several million dollars to enhance or build roads and parks along the banks of the oil-soaked river.

In Mayfield, 4 months later, according to an account in the *New York Times*, "The neighborhood of low-slung brick homes is largely deserted, a ghostly column of empty driveways and darkened windows, the silence broken only by the groan of heavy machinery pawing at the ground as remediation continues." Exxon offered to buy the 22 vacated homes, but only at their post-spill values, which were a fraction of their pre-spill sales prices. Local homeowners were massive losers even as the company cast itself as a benefactor that was paying up for its spilled oil. The company also had spent $2 million on temporary housing for residents and more than $44 million on the cleanup by mid-2013 (Frosch 2013).

7.13 On the Ground in Tar Sands Country

In the midst of the debate over the environmental effects of tar sands, in late August 2013 a large toxic waste spill, the largest of its kind in North American history, rolled over indigenous lands in northern Alberta. The *Toronto Globe and Mail* reported that "the substance is the inky black color of oil, and the treetops are brown.... Across a broad expanse ... the landscape is dead. It has been poisoned by a huge spill of 9.5 million liters of toxic waste from an oil and gas operation in northern Alberta, the third major leak in a region whose residents are now questioning whether enough is being done to maintain ageing energy infrastructure" ("'Every Tree'" 2013).

"Every plant and tree died," Dene Tha' First Nation Chief James Ahnassay told the *Toronto Globe and Mail* ("'Every Tree'" 2013). The Dene said that the spill ran along a trapline (animal-harvesting area) about half a mile from its reserve, a mile from the same people's fishing grounds in the Zama River, where its members fish. The 103-acre spill sprang from a breach in a pipeline owned by the Apache Corporation near Alberta's far northern border. Company representatives said that it was salty water tainted with a small amount of oil and that no people were harmed.

Such a spill is not a one-time event. A 1000-person Beaver Lake indigenous community near Lac La Biche (in northern Alberta) now lives with a moonscape all the time and has been nearly surrounded by oil sands extraction tar pits, as its traditional hunting and fishing range has been laced by 600 miles of tar-covered roads. In 2008, the Beaver Lake Cree Nation filed suit against the governments of Alberta and Canada alleging breach of treaty rights. The defendants fought to have the suit dismissed (as "frivolous") but failed as the Court of the Queen's Bench upheld the community's standing in court in a ruling by Justice Beverley Browne. The governments appealed, and on April 30, 2013, the Alberta Court of Appeal again upheld their right to sue. The suit claims co-management rights under treaty.

7.14 Oil on the Water Table

Brandi Morin of the Indian Country Today Media Network quoted Beaver Lake Cree Nation attorney Drew Mildon as saying that should this case be successful, it could be a precedent-setting impediment to oil sands development, with repercussions throughout at least Alberta, maybe much further. "This is the thing that can keep tar sands development from moving forward," he said (Morin 2013). Mildon said that the effects of oil production through steam-assisted gravity drainage, which is most common in the area, is subtly damaging. "You end up with oil in the water table, with animals being poisoned, with forests being so fractured that it no longer supports any of the fur bearers or other animals on which First Nations depend for food" (Morin 2013). "Science fiction is rife with fantasies of terra-forming—humans traveling to lifeless planets and engineering them into earthlike habitats," wrote Naomi Klein (2014). "The Canadian tar sands are the opposite: terra-deforming. Taking a habitable ecosystem filled with life and engineering it into a moonscape where almost nothing can live" (Klein 2014, 139).

British Columbia photographer Garth Lenz said, "I'd heard about the tar sands but I hadn't been [there], so I went there and spent a couple of days and was pretty much flabbergasted by the scale of the devastation and the impacts. I had photographed industrial devastation all over, including some of the most massive clear-cuts on the planet, right in British Columbia and in Chile and Patagonia, so I'd seen that massive industrialization of the landscape on a huge, huge scale," he said.

"But I was completely unprepared for what I found, because this is just completely off-the-grid crazy—the scale is unbelievable" ("Athabasca Oil Sands" 2011).

Lenz pointed out that the development harms fish and caribou that indigenous peoples in the area rely on. "It's the complete eradication of an ecosystem," Lenz said. "I mean, the forest is clear-cut, the wetlands are drained and dredged, the soil is dug up, replaced by massive mines and toxic ponds which you can see from outer space" ("Athabasca Oil Sands" 2011). Chief Bill Erasmus of the Yellowknife Northwest Territories in northern Canada said that "our people, in some areas, can no longer eat the fish. ... Our people can no longer drink the water. Water levels are decreasing. Where I'm from, it's never been like that before" (Capriccioso 2011).

7.15 Tribes and Nations Forge Alliances

In August 2011, the National Congress of American Indians (NCAI) declared important opposition to the Keystone XL Pipeline that helped to stop it. "Based on the relatively poor environmental record of the first Keystone pipeline, which includes numerous spills, U.S. regulators shut the pipeline down in late May, 2011," said an NCAI resolution, which concluded that "It is probable that further environmental disasters will occur in Indian country if the new pipeline is allowed to be constructed." The Assembly of First Nations of Canada also expressed similar opposition. The NCAI urged the United States to reduce dependence on oil from tar sands that makes such a pipeline necessary with "work towards cleaner, sustainable energy sources" ("NCAI Condemns" 2011). "Homeland and economic security starts with energy security, but Indian Country wants it to be done right," said NCAI President Jefferson Keel, "not at the expense of the health of our communities and resources, both tribal and non-tribal. During challenging economic times in our country and in our tribal nations, domestic energy when developed responsibly can create jobs while ensuring that our people and natural resources remain safe and plentiful" ("NCAI Condemns" 2011).

People from roughly 25 First Nations and tribes in Canada and the United States met on Yankton Sioux (Ihanktonwan) lands for 3 days in late January 2013 to draft a common statement opposing Canadian tar sands and the Keystone XL Pipeline. They called the meeting the Gathering to Protect the Sacred from the Tar Sands and Keystone XL. A signing ceremony was held, along with honoring songs and statements by Chief Arvol Looking Horse, Keeper of the White Buffalo Calf Pipe Bundle and a spiritual leader of the Lakota, Dakota, and Nakota, as well as other notable people, including First Nation actress Tantoo Cardinal (who is Cree) and Debra White Plume (an Oglala Lakota) of the activist group Owe Aku (Bring Back the Way). White Plume and Cardinal had been arrested in front of the White House in 2011 protesting Keystone XL. White Plume also had been arrested in 2012 on Pine Ridge along with others who blockaded trucks carrying loads of machinery headed for the Alberta oil sands mines.

7.15 Tribes and Nations Forge Alliances

"We're here to stand together to protect Mother Earth [for] future generations," Looking Horse said at the treaty meeting. "To us, treaties are the supreme law. We came together with good hearts, good minds." The common statement says that oil sand mining presents "unacceptable risks to the soil, the waters, the air, sacred sites, and our ways of life" (Berry 2013). The alliance's statement asserts that pipelines "carry the possibility of pipeline and tanker oil spills, pose health and ecological threats, and infringe on sacred and historic places, burial grounds and irreplaceable cultural resources" (Berry 2013).

This alliance was forged after an initial meeting in late 2012 when about 20 spiritual leaders of indigenous peoples across North America gathered at a sun dance in South Dakota to compose a declaration "to protect Mother Earth from the impacts of development in Canada's oil sands mega-project." Mohawk, Lakota, Navajo, Apache, Aztec, and Ojibwe leaders took part, hosted by Sicangu Leonard Crow Dog, Lakota spiritual leader. Sun dancers who had come from an area spanning Alaska to Peru pledged their support as well. The idea for the statement originated with Sundance Chief Rueben George of the Tsleil Waututh Nation in North Vancouver, British Columbia. Crow Dog, who has been an American Indian spiritual adviser since the 1970s, said, "It is our responsibility to protect and care for these elements in accordance with our own sacred laws and traditions, and in doing so, maintain our spiritual relationship to the land, water, plants, animals, our ancestors, all of our relations and future generations" (Ball 2012b).

A declaration by the spiritual leaders said, in part, "We therefore will defend our land and exercise our own laws and traditions from all directions to oppose all tar sands pipeline and tanker projects, and the tar sands development itself, all of which threaten the physical, mental, emotional and spiritual well-being of all of our relations" (Ball 2012b). "The Keystone pipeline goes right through their [Lakota] territory," George told the Indian Country Today Media Network. "It was beautiful to have so many chiefs come together, because [be it] in Peru or up in Alaska or here in Vancouver, we're dealing with the same problems: fossil fuel problems. We also spoke of alternatives, too. Our ceremonies and beliefs come from the elements of Earth: from fire, earth, water and sky. That's why we believe our lands and waters are sacred" (Ball 2012b).

On October 1, shortly before the spiritual leaders issued their statement, the Athabasca Chipewyan First Nation (ACFN), whose lands are situated near the oil sands mines, issued a constitutional challenge based on Treaty 8 asserting that Alberta's government and Shell Canada had violated their legal rights by failing to consult them regarding the development of the mines. The Chipewyan demanded "mandatory consultation with First Nation treaty signatories. It also demanded observance of their right to access their territories for hunting, trapping, harvesting and other traditional uses. The Chipewyan alleged that the Jackpine expansion, by damaging the environment, would prevent them from engaging in those activities and resources and thus violate the treaty" (Ball 2012a). Treaty 8 guarantees signatories a right to self-governance and use of their lands and resources for sustenance. ACFN spokesperson Eriel Deranger said,

> What are the costs of pushing the industry through? We're talking about doubling production in the tar sands. We're already having problems with the current pace of development. Doubling it is psychotic. Some people think the tar sands and First Nations people can coexist, [but] I don't know how you could possibly rip up thousands of kilometers of boreal forest and traditional territories, de-water, poison and contaminate river systems, and consider that a plausible way for coexistence? (Ball 2012a)

In 2014, a report by the Joint Oil Sands Monitoring Program alleged that dams holding back wastewater from oil sands mining waste were seeping 6.5 million liters (or 1.7 million gallons) daily into the Athabasca River's groundwater, including bitumen, arsenic, cadmium cyanide, phenols, naphthenic acids, and other metals. "In short, [the study] highlights past studies identifying tailings ponds as significant sources of groundwater contamination, and brings to light that groundwater contaminated by leaking tailings ponds is almost certainly flowing into the Athabasca River," said William Donahue, a freshwater science specialist based in Edmonton (Nikiforuk 2014).

Shell Canada was proposing to increase oil sand production by 100,000 barrels a day as part of the Jackpine project in a 31,429-square-mile "disturbance area" that indigenous peoples use for hunting, trapping, and fishing in the Muskeg River area. Shell spokesperson David Williams said that the company had consulted with the First Nations repeatedly and that it does respect ACFN's treaty rights. "Shell has engaged extensively with the Athabasca Chipewyan for more than 15 years," he told the *Toronto Globe and Mail*. Deranger said that mere contact did not constitute consultation and that meeting solely with Native people who stood to benefit from the mining did not meet the terms of the treaty.

"There are First Nations [people] who think the tar sands are great," Deranger admitted. "People have jobs. People now can afford to take their kids to Edmonton to go to the dentist. These are luxuries for people. But we have to start weighing the costs … [including] hundreds of toxic tailings [waste] ponds, open-pit mines, significant emissions and polluted rivers across a giant swathe of Alberta," Deranger said (Ball 2012a). Richard Ray Whitman of the Yuchi/Muskogee Creek tribes said that many American Indians in the United States support Canadian First Nations that have been affected by pipeline development. "Our concerns also go far beyond what might happen here," he said. "Because of the extraction of the tar sands, we are currently witnessing the devastation of lands considered sacred by indigenous people in Canada. Opposing the Keystone XL Pipeline means standing in solidarity with all our Native brothers and sisters in the Northern U.S. and Canada" (Capriccioso 2011).

7.16 Native Peoples Unite against the Keystone XL

Delegations of Native peoples have visited the State Department and the White House as well as the E.P.A., expressing their opposition to the Keystone XL and other pipelines that are planned to cross over or close to their lands. In May 2013, Ten U.S. Native tribes and nations in the pipeline's proposed line of transit declared their opposition:

> On this historic day of May 16, 2013, ten sovereign Indigenous nations maintain that the proposed TransCanada/Keystone XL pipeline does not serve the national interest and in fact would be detrimental not only to the collected sovereigns but all future generations on planet Earth. This morning the following sovereigns informed the Department of State Tribal Consultation effort at the Hilton Garden Inn in Rapid City, SD, that the gathering was not recognized as a valid consultation on a "nation to nation" level ("Full Text" 2013).

The statement was signed by the Southern Ponca, Pawnee, Nez Perce Nation, Oceti Sakowin (Seven Council Fires People), Sisseton-Wahpeton Oyate, Ihanktonwan Dakota (Yankton Sioux), Rosebud Sioux, Oglala Sioux, Standing Rock, Lower Brule Sioux, Cheyenne River Sioux, and Crow Creek Sioux. At about the same time, the Great Plains Tribal Chairman's Association (which includes 16 tribal presidents in Nebraska, South Dakota, and North Dakota) also supported this position, which expresses concern regarding "potential pipeline impacts on natural resources, especially our water: potential spills and leaks, groundwater and surface water contamination," as well as climate change made worse by the exploitation of fossil fuels. "America would be better served," the common statement said, "by a comprehensive program to reduce its reliance on oil, and to invest in the development and deployment of sustainable energy technologies, such as electric vehicles that are charged using solar and wind power." The statement also opposed the pipeline on treaty-rights grounds: "If the Keystone XL pipeline is allowed to be built, TransCanada, a Canadian corporation, would be occupying sacred treaty lands reserved in the 1851 and 1868 Fort Laramie Treaties, [which were negotiated by representatives of the United States with those of several indigenous nations]. It will be stopped by unified resistance" ("Full Text" 2013).

On May 16, 2013, elders and chiefs of at least 10 Native American nations walked out of a meeting with U.S. State Department officials in Rapid City, South Dakota, on May 16 in which the government was attempting to engage in tribal consultation over the Keystone XL Pipeline. Many in the group that earlier had signed a joint statement opposing the Keystone XL (Southern Ponca of Oklahoma, Pawnee Nation, Nez Perce Nation, Sisseton-Wahpeton Oyate, Ihanktonwan Dakota Yankton Sioux, Rosebud Sioux, Oglala Sioux, Standing Rock, Cheyenne River

Sioux, and Crow Creek Sioux) called the meeting "invalid" and asked to meet directly with President Obama on pipeline-related issues.

The Native leaders who walked out of the meeting with State Department officials issued a statement objecting to the pipeline for several reasons. "On this historic day of May 16, 2013, sovereign Indigenous nations maintain that the proposed Trans-Canada/Keystone XL pipeline does not serve the national interest and in fact would be detrimental not only to the collected sovereigns but all future generations on planet Earth. This morning the following sovereigns informed the Department of State Tribal Consultation effort at the Hilton Garden Inn in Rapid City, SD, that the gathering was not recognized as a valid consultation on a 'nation to nation' level.... Eventually all remaining tribal representatives and Tribal Historic Preservation Officers left the meeting at the direct urging of the grassroots organization Owe Aku," the chiefs said in their statement. "Owe Aku, Moccasins on the Ground, and Protect the Sacred are preparing communities to resist the Keystone XL pipeline through Keystone Blockade Training" ("Chiefs" 2013).

The NCAI kept a detailed watch on the State Department's handling of the Keystone XL issue, severely criticizing sections of its report on oil spill prevention and remediation as well as its lack of detail on American Indians' and Alaska Natives' water quality and supply. "In total," the NCAI statement concluded, "if these concerns are not addressed sufficiently or mitigated to the fullest extent, it is in the best interest of the United States to reject the Keystone XL pipeline permit solely on the basis of the federal trust responsibility to tribal nations.... The project as outlined in the DSEIS poses tremendous risks to the cultural and natural resources of tribal nations and is not in the best interest of the tribal nations and their citizens" ("Fill Gaps" 2013).

7.17 A Blockade at Pine Ridge

The Oglala Sioux on the Pine Ridge Indian reservation expressed concern that the path of the Keystone XL would cross the route of a pipeline delivering water to them as well as the neighboring Rosebud Sioux in two places (Woodard 2013). A blockade at Pine Ridge began on March 5, 2012, after a resident of Wanblee on the reservation was forced off the road by a convoy of 150-ton semitrailers. She called several neighbors, who converged on the site, blockading the road with their cars and stopping the TOTRAN Transportation Services rigs. Traffic stopped for 6 hours as about 40 cars joined the blockade, containing between 50 and 75 people. Tribal police arrested five people who refused to move (Alex White Plume, Debra White Plume, Sam Long Black Cat, Andrew Iron Shell, and Don Iron Shell) on charges of disorderly conduct.

The trucks' drivers said they were destined for Canadian mines, hauling "treater vessels," which use intense heat to separate oil and gas from other substances, invoiced at almost $260,000 per load. The *Rapid City Journal* reported the next day that TransCanada denied the shipments were part of Keystone XL construction.

However, Oglala Nation Vice President Tom Poor Bear said that South Dakota Gov. Dennis Daugaard's secretary of tribal relations, J. R. LaPlante, had told him the trucks had been routed through Pine Ridge by TransCanada to avoid paying $50,000 per vehicle in fees for using South Dakota highways. "He said, 'Mr. Poor Bear, I want to apologize. The South Dakota Department of Transportation'—and then he named a couple senators and himself—'had a meeting a couple weeks ago to reroute these trucks that are holding these pipes and water tanks that are going to Canada for the Keystone pipeline. ... We had to reroute them through your reservation.'"

Poor Bear then replied, "Mr. LaPlante; I took a lead in opposing the Keystone pipeline, as well as our neighbors the Rosebud Sioux Tribe, and said that we oppose the pipeline and you allowed them to come through our reservation without asking permission?" (Schilling 2012). Poor Bear had called out to President Obama as he spoke to university students in Denver during October 2012 urging him to disapprove a permit for the $7 billion, 1711-mile Keystone XL Pipeline. The president replied, "I hear you," and told the delegation that "no decision has been made" (Berry 2013). Speaking for the Oglala Sioux delegation, Poor Bear said, "The fight against this pipeline is far from over; we must become as one to protect our mother the Earth, and future generations," urging "on behalf of the Oglala Lakota" for "each and every one of our relatives from other tribes and nations to stand with us in unity and solidarity to protect what is sacred." The Oglala Sioux Tribal Council also passed a resolution against the pipeline, saying that it would involve "accessing a 300-foot-wide corridor through unceded treaty lands of the Great Sioux Nation" as included in the Fort Laramie treaties of 1851 and 1868 (Berry 2013).

7.18 Another Blockade by the Nez Perce

The Nez Perce hosted Lewis and Clark in 1805 and later, in 1877, led the U.S. Army on a months-long chase over some of the roughest landscape in North America. In 2013, the fighting spirit of the Nez Perce again was engaged by transport through their lands of huge loads destined for Alberta's tar sands fields. On August 5, 2013, more than 200 Nez Perce intercepted a convoy of trucks destined for the oil sands fields of Alberta, forming a blockade that was broken up by police, who arrested 30 people, including chairman Silas Whitman and six members of the tribe's executive council.

In 2013, as in 1877, the Nez Perce did not appreciate being pushed around, and they quickly put events into context. "The development of American corporate society has always been—and it's true throughout the world—on the backs of those who are oppressed, repressed or depressed," Silas Whitman, chairman of the Tribal Executive Committee, told the *New York Times*. "We couldn't turn the cheek anymore." After their meeting, the Nez Perce leaders decided to face arrest as a group (Johnson 2013). Whitman, 72 years of age at the time, was one of several Nez Perce arrested at the barricade as police dismantled it. The blockade lasted four nights.

"Lights! Lights!" they shouted as amber-colored flashers on a pilot truck came into view in the darkness just after midnight along Highway 12 on the Nez Perce reservation. The Indian Country Media Network reported that "a rush of more than 200 tribal members and others—from grandmothers to children—followed, all determined to stop a football-field-sized mega-load from passing through their sacred lands," erecting wooden barriers, as police cruisers idling in the darkness nearby flipped on their headlights and rooftop light bars and rolled to within yards of the blockade.

It was an eerie blue-and-red strobe-lit standoff. Tribal members "sang and drummed and whooped, while state and tribal police faced them, arms crossed. Looming behind the police came the mega-load, its cylindrical face appearing as an enormous ghostly moon, swathed in a white tarp" (Taylor 2013). The trucks they stopped were carrying a 23-foot-tall, 322-ton water evaporator hooked onto a diesel truck front and back, 243 feet long.

Early in 2013, the Nez Perce Executive Council adopted a position opposing transportation of very large (megaload) mining equipment through its land along 780 miles of Highway 12. Such shipments have become common there. "They will also be traveling really close to our creation story," said McCoy Oatman, chairman of the Nez Perce Executive Committee. "It's called 'The Heart of the Monster.' It's essentially the birthplace of the Nez Perce people. It [the route] runs dangerously close to that site, and that's pretty significant for us" ("Nez Perce Victory" 2013).

The protests paid off. The Nez Perce also joined with Idaho Rivers United, an environmental group, and took the matter to Boise's federal district court, where Chief Judge B. Lynn Winmill stopped further transport of megaloads until the Nez Perce and the U.S. Forest Service assessed their environmental impact. The Nez Perce believe that their scenic route was being turned into an industrial corridor as federal and state officials ignored them. In September 2013, the Forest Service closed the Nez Perce route to megaloads and required consultation with the tribe in the future. "The closure order is effective immediately and is in place until the agency agrees to lift it, mostly likely when forest officials complete a study on the impacts the shipments could have on the river corridor," the Associated Press said. "The agency has also vowed to consult with tribal officials on the impact to treaty rights and cultural values" ("Nez Perce Victory" 2013).

7.19 Protests Spread

Protests of the megaloads soon spread beyond the Lakota and Nez Perce. The Ho-Chunk (Winnebago) Tribal Council in 2012 adopted a resolution against all tar sand mining on their lands. The sand was to be used for fracking. Such declarations spread across North America, as Native Americans took a leading role in resisting tar-sands mining. Native protesters pointed out that water is used in very large amounts at high pressure to fracture shales, releasing the gas, in combination with a "fracking cocktail [that] includes acids, detergents and poisons that are not

7.19 Protests Spread

regulated by federal laws but can be problematic if they seep into drinking water" (Brantley and Meyendorff 2013). Methane release can create the risk of explosion.

On August 9, 2013, about 150 people, Native and non-Native, arrived at Pier 96 in New York City, having paddled as far as 300 miles along the Hudson River and its tributaries, in defense of the Two Row Wampum, a mutual respect treaty first negotiated in the early seventeenth century between the Haudenosaunee (Iroquois) and immigrating Europeans. They docked their canoes and kayaks, then hiked to the UN headquarters building on the eastern side of Manhattan Island. Their purpose was both historical and environmental, and one of their major targets was fracking, a call that was renewed on January 8, 2014, at Governor Andrew Cuomo's State of the State Address in Albany ("Call to Action" 2014).

Riders undertook a 150-mile horseback journey from the Pine Ridge reservation to the Cheyenne River reservation, both in South Dakota, near the formerly proposed route of the proposed Keystone XL Pipeline. "Thundering across the plains on horseback, along the routes of two proposed oil pipelines, Earth's Army has wound up its journey to draw attention to not just TransCanada's Keystone XL pipeline, but also a lesser-known one being proposed by Enbridge across White Earth territory," said one eyewitness account. The ride was led by Percy White Plume, whose ancestors survived the 1890 Wounded Knee massacre. "We can drink bottled water, but our relatives in the horse nation, the buffalo nation and the [other] animals cannot drink bottled water, our water is sacred," he said ("Anishinaabe" 2013). The riders were organized by the Horse Spirit Society of Wounded Knee and supported by the Swift Family Foundation, 350.org, the U.S. Climate Action Network, and Honor the Earth.

At about the same time, nine people were arrested on June 24, 2013, after they chained themselves to a work trailer and an excavator in Seminole, Oklahoma, in protest of the Keystone XL, blocking construction of a pump station for the pipeline on Seminole treaty land. Halfway across North America, the Columbia River Inter-Tribal Fish Commission joined the opposition. "The development, transportation, and use of Canada tar sands oil will have long-lasting negative effects and pose significant threats to the Columbia River Basin, its natural resources and the people who reside there," stated a resolution passed on August 23, 2013, by the Umatilla, Yakama, Warm Springs, and the Nez Perce, who had been protesting the megaload transports through their lands for 3 years by 2013.

Native people at Umatilla and the Confederated Tribes of Warm Springs in northern Oregon also protested the passage through their lands of a 45-ton megaload (18 feet tall, 22 feet wide, and 376 feet long) bound for Alberta's tar sands fields in early December 2013. About 70 people carrying signs reading "Hands Off Our Planet! Stop the Tar Sands" swarmed the megaload and impeded its path. "This has gone too far. Our children are going to die from this," some of them said in an online video posted on the Internet by Portland Rising Tide. "If we don't stop this now ... [future generations] are going to ask, 'Where were you? Where was our tribe?'" (Dadigan 2013). A few locked their bodies to the huge truck. The megaload forced the protesters aside but was later stopped by snow and ice near Pendleton, Oregon.

Several activists were arrested on charges of disorderly conduct after police removed the locks.

Alberta's Lubicon Lake Cree Nation protested fracking as well as tar sands mining on and near their lands, the latter with a blockade on unceded territory of a road northeast of Peace River in late 2013. The Calgary-based oil and gas company Penn West Petroleum Ltd., whose site was being dismantled, filed suit to dismantle the Lubicons' blockade. "The judge denied [us] the opportunity to raise any of the constitutional issues and arguments for the Lubicon," said Garrett Tomlinson with the Lubicon Lake Nation. "More time must be provided for both sides to be heard" (Troian 2014). Some of the Lubicon argued that the lack of a treaty between their tribe and the Canadian government rendered any permits to Penn West null and void. The Lubicon Cree leadership itself was split, however, between a faction supporting Penn West and one opposing it.

The Lubicon Lake Nation's leadership argued that Canada had never entered into a treaty with them. They asserted that lack of a treaty rendered permits for oil and gas development on Lubicon lands issued to Penn West by Alberta null and void. Penn West had explored fracking between Haig and Swan Lakes, where Lubicon Lake people carry out traditional activities such as fishing and hunting. "We have to stand up for Mother Earth." they said.

7.20 More Protests, More Arrests

"We have to stand up for our sacred water—for our children, our grandchildren, for the coming generations," said Lakota activist Debra White Plume as she was arrested at the White House on September 2, 2011, with 186 others protesting the Keystone XL Pipeline. Several of those arrested were Canadian Native people who also were protesting oil sands development on or near their homelands. White Plume said that leaks from the pipeline could desecrate the freshwater Ogallala Aquifer near her homeland in Pine Ridge, South Dakota, a violation of the 1868 Fort Laramie Treaty. "It is with great honor that I come here today to ask President Obama to stand with us for Mother Earth against Father Greed," Plume said (Capriccioso 2011).

On April 22, 2014, cowboys and Indians united in opposition to the Keystone XL on the Mall in Washington, D.C., many of them on horseback, to the sound of beating drums, as about 100 people gathered around a large tipi. The demonstration was assembled by the Cowboy and Indian Alliance, representing landowners in Nebraska and Native opponents of the pipeline who marched and rode along Independence Avenue onto the Mall.

When 68-year-old Canadian folksinger Neil Young devoted a tour to help the ACFN's legal campaign to impede expansion of oil sands mining, Prime Minister Stephen Harper said that his "rock-star lifestyle made him unfit" to criticize it. In response, Young took the stage at a concert in Toronto with an audience of 2700 on January 12, 2014, with song lyrics needling Harper as a supporter of genocide. David Ball (2014a) of the Indian Country Today Media Network wrote that "the

7.20 More Protests, More Arrests 171

legendary Canadian musician substituted Harper's name into his song 'Pocahontas,' which describes massacres of Indigenous peoples." The Chipewyan First Nation said that contamination of soil and water from tar sands mining was causing cancer rates among their people to rise rapidly. "In the street outside Massey Hall," wrote Bell, "dozens held a round dance—a fixture of the Idle No More movement that has swept the country over the past year." (Ball 2014a).

In addition to Toronto, Young's "Honour the Treaties" tour visited Winnipeg, Regina, and Calgary. "We made a deal with these people," Young said. "We are breaking our promise. We are killing these people. The blood of these people will be on modern Canada's hands, and it will be the result of not just a slow thing, but of a fast and horrific thing if this continues. Believe me, these people are not going to sit back and let modern Canada roll over them" (Ball 2014a).

Leonardo DiCaprio, a celebrity actor and filmmaker, arrived in tar sands country in Fort Chipewyan, Alberta, and met with members of the ACFN to discuss creating a documentary on the Alberta environmental situation. On August 22, 2014, DiCaprio met with "Black Swan" director Darren Aronofsky, both of whom received a blast of invective from a spokesman for the Canadian Association of Petroleum Producers, wrapped in *Canadianismo:* "Like Canadians, we are growing tired of the fad of celebrity environmentalists coming into the region for a few hours or a few days, and offering their ideas and solutions to developing this resource" (Leo DiCaprio 2014).

President Obama was met with protests of the XL Pipeline when he became only the fourth sitting U.S. president to visit an Indian reservation on June 13, 2014, at Pine Ridge. "President Obama must reject this pipeline and protect our sacred land and water ... [it] is a death warrant for our people," said Bryan Brewer, president of the Oglala Sioux tribe, as the Indigenous Environmental Network demonstrated against the project. Obama's remarks at a powwow on the Standing Rock Sioux reservation in central South Dakota focused on increasing economic opportunities for Native Americans.

In early 2015, the Fort Peck reservation's Sioux and Assiniboine peoples of Montana passed a resolution opposing the Keystone XL, saying that "the pipeline would jeopardize drinking water projects for the reservation." The pipeline, as proposed, would have crossed watercourses upstream from the Fort Peck Dam, sites from which a rupture could contaminate the entire Missouri River water supply. This route includes the intake for the Assiniboine/Sioux Water Project on the Fort Peck reservation. "This project will provide water for at least 30,000 people," said Bill Whitehead, a tribal member, former councilman, and former state legislator who was chairman of the Assiniboine/Sioux Water Project. "Through our treaty rights and water compacts it covers all of northeastern Montana, including the reservation and four counties" (McNeel 2015).

Native Americans, the largest minority group in South Dakota at 9% of the population, were among the most vocal opponents of the pipeline; 313 miles of its proposed route would have run through the state. Some Rosebud Sioux had maintained a "spirit camp" opposing the pipeline for more than a year by mid-2015 (with a trailer and a tent but without running water) on tribal land near the anticipated Keystone XL route, where, according to a report in the *New York Times*, they have "vowed to use the site as a base camp for protesters if construction ever begins. Several activists, said they would risk arrest through civil disobedience" (Smith 2015). Faith Spotted Eagle, the chairwoman of the Yankton Sioux tribe's Treaty Council, emerged as a leader of an increasingly organized coalition of Native Americans, landowners, and grassroots groups seeking to block construction of the Keystone XL across South Dakota.

7.21 Invoking the Fort Laramie Treaties

Invoking Article I of the 1868 Fort Laramie Treaty, the Lower Brule Sioux tribe voted during April 2015 to reject the Keystone XL Pipeline and evict TransCanada from its lands "in direct response to the unethical business practices that Trans-Canada has demonstrated over the last six years." Acting Chairman Kevin Wright said that "as descendants of the people of this land we have witnessed destruction of many magnitudes, we are concerned for our land, water, and most importantly not only the physical wellbeing of our people but spiritual wellbeing as well. I am first a human being, not a politician, when it comes to these matters. I believe in protecting our people and look to more ecological ways of living" ("Lower Brule" 2015). The treaty allows the eviction of "bad men" who do not act in good faith. Sioux land under the 1868 treaty included about 40% of present-day South Dakota, including part of the Keystone XL's proposed route.

Following a commencement speech in May 2015 at the Lake Area Technical Institute in Watertown, South Dakota, President Obama accepted a star quilt created by Sisseton-Wahpeton Dakota artist DeVon Burshiem, who had embroidered "NOKXL" ["No Keystone XL") on its reverse side. "*Tasina Pezuta Win emakiyapi ye* ["They call me medicine blanket woman"]. "It was truly an honor and a blessing to know that the quilt I was making would soon be in the hands of the President of the United States, Barack Obama!" said Burshiem by way of the Indigenous Environmental Network. "Heartfelt prayers were made during that time, tears ran down my face as I prayed for our land and for our water, *mni wicozani* [water is life], without water we are without life. #NOKXL" ("Message" 2015).

7.22 Shipping Oil: And Spilling It—By Rail

Lacking approval for the Keystone XL and other pipelines, oil companies took to shipping their product by train, 100 tank cars or more at a time, leading in some cases to spectacular wrecks and spills (such as in North Dakota and Quebec in 2013). More than 400,000 tank cars of crude were shipped in 2013, up from 10,000 in 2008. Two-thirds of the Bakken formation's shale oil departed by train in 2013. At the same time, the number of gallons of oil spilled by railroad in 2013 exceeded that of the previous 37 years (1975–2012) combined, according to the Association of American Railroads (Krauss and Mouawad 2014, A-1). Canadian oil companies added to the rail network, which, by 2015, was capable of carrying just as much oil west or south to new terminals in the United States, China, and other world ports. Native resistance also keyed on the growing rail network, with mobilizations among the Quinault (on Washington State's Pacific Coast near one proposed terminal site) and others. The amount of crude oil shipped by rail in the United States increased from 9500 barrels in 2006 to 66,000 in 2011 and 234,000 in 2012 ("Lac-Mégantic" 2013).

"Over all," reported the *New York Times*, "Canada is poised to quadruple its rail-loading capacity over the next few years to as much as 900,000 barrels a day, up from 180,000 today ... despite a derailment in the lakeside Quebec town of Lac-Mégantic in July [2013], in which a runaway oil train bound for a refinery in eastern Canada exploded, killing dozens of people and bankrupting the railway company" (Krauss 2013). "They don't give up," Jesse Prentice-Dunn, a Sierra Club policy analyst, said of the oil industry. Rail shipment is about $5 per barrel more expensive than pipelines. "Several Washington and Oregon refiners and ports are planning or building rail projects for Canadian heavy crude as well as light oil from North Dakota. The Texas refinery giant Tesoro and the oil services company Savage have announced a joint venture to build a $100 million, 42-acre oil-handling plant in the Port of Vancouver on the Columbia River that could handle 380,000 barrels of oil each day if permits are granted" (Krauss 2013).

7.23 The Quinault Resist Oil Transport

After an oil train explosion killed 47 people on July 6, 2013, in Lac-Mégantic, Quebec, destroying the center of the town of 6000 people, the Quinaults on the Pacific Coast of Washington State restated their opposition to the transport of oil and coal in their verdantly forested homeland. "It could have easily been Hoquiam" (a town near the Quinault Nation on the Pacific coast of Washington State), said Fawn Sharp, Quinault council president. "It is not a matter of 'if' these shipments will cause a major spill; it's a matter of 'when,'" said Sharp ("Lac-Mégantic" 2013). The Quinault oppose plans by Westway Terminal Company from Louisiana and Texas to construct a terminal for the shipment and storage of 800,000 barrels of oil in Grays Harbor near Hoquiam. The company's plans call for shipping and storing 10 million barrels of crude oil in that area.

"The massive train, oil barge and ship traffic this project will bring to Grays Harbor is a tragedy waiting to happen," Sharp said. "There will be spills and they will harm salmon, shellfish, and aquatic life, trample our treaty rights and cultural historic sites, and tie up traffic for extensive distances." Sharp added that the facility would violate Quinault treaty rights and their standards of "good stewardship and common sense.... The risk is not worth a few more, unsustainable jobs. Far too much is at stake, and there is simply no way oil train proponents can pass the straight face test and tell us that their proposal is safe. Lives are at stake—fish and wildlife resources, water quality, and much, much more. These are the same type of rail cars that will come pouring through our area, and unquestionably threaten the lives and safety of our people and resources" ("Lac-Mégantic" 2013).

7.24 More Resistance to Fracking

Fracking sparked contention across Indian Country, from eastern New York State and Pennsylvania to New Brunswick, Canada, where 12 protesters were arrested on June 21, 2013, on National Aboriginal Day for blocking work crews. Royal Canadian Mounted Police (RCMP) Corporal Chantal Farrah said, "They were attempting to block the heavy equipment from traveling on the road.... Now the people ... were informed that they were breaking the law and that they needed to move. They refused, and 12 people were arrested ... seven men and five women" (Ball 2013b).

Native Americans are not opposing fracking alone, of course. By November 2013, more than 100 cities and towns in the United States had passed laws against fracking, according to a nonprofit monitor, FracTracker. Brantley and Meyendorff (2013) reported in the *New York Times* that "opposition to fracking has been considerable, if not unanimous, in the global green community and in Europe, in particular. France and Bulgaria, with the largest shale-gas reserves in Europe, already had banned fracking." Protesters blockaded some fracking test sites in the United Kingdom and Poland. In Britain, near Balcombe in West Sussex, as many as 1000 demonstrators set up a tent camp during the summer of 2013 to protest test drilling by the energy company Cuadrilla in what became a symbol of opposition to fracking. More than 100 people were arrested, including a member of Parliament from the Green Party, Caroline Lucas.

"The company removed its test rig and left the site in late September," wrote *The Times'* Steven Erlanger (2013).

In 2011, women of Alberta's Blood Tribe set up a blockade against fracking on their lands to publicize their opposition. "As a grandmother and a mother, and as a member of the Blood Tribe, I have a vested interest like everyone that shares Mother Earth with us to help protect what we have," explained Cassie Brewer, holding up a sign decrying the practice. "The treaties pushed us here. We don't have another land to go to" (Ball 2013b).

Another fracking blockade was set up on the Elsipogtog First Nation of New Brunswick. On October 1, 2013, members and leaders of the Elsipogtog Nation observed Mi'kmaq Treaty Day in Rexton on Provincial Route 134. Usually a day for emphasis on Mi'kmaq history and culture stemming from their 1752 pact with the British Crown, Treaty Day became an emphatic protest against fracking. Several dozen protesters blocked a road used by vehicles from SWN Resources Canada, a subsidiary of the Southwestern Energy Company, based in Houston, Texas. During the spring of 2012, SWN had begun seismic testing near the Elsipogtog First Nation. "Right now we're standing our ground and asking [SWN] to pull their equipment out

7.24 More Resistance to Fracking

of New Brunswick so that it will resolve peacefully," said John Levi, the warrior chief of Elsipogtog First Nation. "We're not going to back down" (Troian 2013a).

The blockade's members issued an "eviction notice" to prospective frackers from Texas. Chief Arren Sock of the Elsipogtog Nation served the notice, as a Band Council resolution: "Whereas Prime Minister Harper and the Canadian Government have washed their hands with regards to the environmental protection of our lands and waters," read Chief Sock from a prepared statement that also described lands which "have been assaulted by clear-cutting" as well as fracking, "We have lost all confidence in governments for the safekeeping of our lands held in trust by the British Crown." Their notice of eviction was "totally ignored by the provincial government and Southwestern Energy," said Sock. "Thus, First Nations peoples "have been compelled to act and save our water, land and animals from ruin," so the "council of Elsipogtog are reclaiming all unoccupied reserved native lands...and put in the trust of our people.... I want a moratorium on fracking in New Brunswick... until there is such a time that they can come up with a safer solution on how to extract shale gas. Right now, I am here for our children and their children's children. Right now nobody can guarantee a safe and effective way of extracting shale gas without harm to the environment.... No more negotiations with anybody" ("Elsipogtog Chief" 2013).

In the early morning of October 17, 2013, the RCMP and local police clad in riot gear broke down the week-old barricade, firing tear gas and rubber bullets and arresting at least 40 people. The attack came after several days of peace. "For numerous days prior, RCMP were allowing the first walking traffic, then one lane of automobile traffic, to pass freely through the blockaded area," wrote Miles Howe for the Halifax Media Co-op. "Anti-shale activists, as a measure of good faith, and in deference to emergency vehicles in particular, had days earlier removed two felled trees that had completely blocked off vehicular traffic" (Howe 2013).

Vincent Schilling (2013b), writing for the Indian Country Today Media Network, described the scene:

> Chaotic: heavily armed Royal Canadian Mounted Police (RCMP) pouring into an encampment of sleeping protesters, leading dogs and carrying assault rifles. Amid burning police cars, pepper-spray-spewing hoses and barking police dogs, 28-year-old Amanda Polchies dropped to her knees, brandishing the only "weapon" she had an eagle feather. Holding it aloft, she began to pray. An iPhone photo of Polchies was snapped, her feather aloft, in front of a wall of Royal Canadian Mounted Police, which went viral on social media, becoming an emblem of the protest.

"I saw elders getting doused with pepper spray. I also saw people's bruises from weapon-fired beanbags and rubber bullets, and police dogs and trucks all over the place. The dogs were barking so loud and jumping that these vehicles were shaking," said Michelin, an Inuit of mixed heritage from Labrador (Schilling 2013a). Some protesters, one of whom was pregnant (as well as a number of elders), faced off with private security guards and members of the RCMP who removed their blockade. Lorraine Clair, 44, according to one eyewitness, "was struck by an RCMP vehicle during an incident. She was also injured during an earlier confrontation with the RCMP," according to the Indian Country Today Media Network. "She's a fish plant worker, so her wrists were already weakened, and she was still healing from that, and then when they arrested her they broke her wrist," said protester Willie Nolan (Troian 2013b).

Howe, who was among those arrested, reported that "the ditch opposite me was already filled with 20-odd police in tactical blue uniforms, pistols already drawn. Three police officers dressed in full camouflage, one with a short-chained German Shepherd, were also near the ditch" (Howe 2013). "In the pre-dawn dark," wrote Howe, "about seven Molotov cocktails flew out of the woods opposite the police line stationed in the ditch. I cannot verify who threw these cocktails. They were—if it matters—lobbed ineffectively at the line of police and merely splashed small lines of fire across the road. A lawn chair caught fire from one cocktail. Two camouflaged officers then pumped three rounds of rubber bullets...into the woods." About 10 minutes after that, the confrontation turned into a melee after two officers served an injunction on Seven Bernard, who was acting as a spokesman for the Warriors at the blockade, standing near a wall of about 60 officers with guns drawn. As Sgt. Rick Bernard (no relation to Seven Bernard) began to serve the paper, Seven Bernard walked away. "Sergeant Bernard threw the injunction at his namesake, saying. 'Consider yourself served'" (Howe 2013).

Mi'kmaq Suzanne Patles attempted to intervene with a peace offering of tobacco as officers began hauling people away in handcuffs under a barrage of rocks and bottles. "Skirmishes then broke out in every direction," wrote Howe (2013). "From the highway side, District War Chief Jason Augustine was being chased by numerous police. In front of me, everywhere really, Warriors were being taken down by numerous RCMP officers in various clothes. Rubber bullet shots were fired by the RCMP, and both Jim Pictou and Aaron Francis claimed that they were hit—in the back and leg respectively" (Howe 2013). The Indian Country Today Media Network reported that "six RCMP vehicles were destroyed by fire, and several improvised explosive devices were discovered and defused," the RCMP said. "These explosives contained shrapnel and had the potential to seriously injure or kill people" ("Molotov" 2013). Some of the blockade members threw rocks and bottles at the police and the RCMP.

The Indian Country Today Media Network reported that "amateur photos and video have appeared online showing heavily armed police on the site and what appear to be snipers in nearby fields and forests. There are also photos of several police vehicles on fire. Those arrested, which may include elders conducting ceremonies at the site, are being detained by the Royal Canadian Mounted Police (RCMP). There were also some reports of shots fired" (Troian 2013b).

The RCMP said that the intervention and arrests were justified because the blockade had become a threat to public safety, while Canadian Assembly of First Nations National Chief Shawn Atleo said in a letter to New Brunswick Premier

David Alward that a peaceful protest had been "disrupted through police and military intervention" (Stone 2013). News of the event spread rapidly on social media, and supporters began to arrive, some from as far away as Alberta. Floyd Augustine, a leader of the Mi'kmaq Warrior Society, issued a call for support. Demonstrations were reported in Montreal, Vancouver, Ottawa, Iqaluit, Nunavut, Minneapolis, New York City, and San Francisco over the next few days as the Sierra Club announced support.

In an interview on Native Trailblazers Radio, Michelin said, "One of the hardest things for me to see was the treatment of Amy Sock [a member of the Elsipogtog First Nation]. She has a background in law and lives in Elsipogtog. This woman is one of the most peaceful people I have ever met in my life. There was a line of police standing around cars, and she came up with a white flag. She said she needed to see her people and wanted to make sure they were okay. She tried to break through the police line and started running. I saw people running behind her. She is about 125 pounds. I saw three big police officers tackle her to the ground like football players. They had her hands zip-tied, [pushing] her to the ground with her knee in her back, and the police officer looked up the same way a hunter poses over his kill. It broke my heart" (Schilling 2013a).

"We're not going to stop," said Sock. "We don't want shale gas to come to New Brunswick.... We have a big nuclear plant here. Once fracking goes on—once we start getting earthquakes—I'm afraid that thing is going to blow up. That's one of my fears. Our priority is Mother Earth." Photos of Sock spread rapidly over social media as she was brutally arrested during the blockade. "I can cry, get mad, pray, forgive, and do my pipe ceremony to keep my strength and courage up," Sock said, "but this is one tough battle. We live by the river.... Our regular diet—fish, clams, eels, bass, salmon, and lobster—that's what we eat. I want to continue eating that without getting sick [from water pollution associated with fracking]."

7.25 Fracking and Earthquakes

Until recent years, earthquakes were very rare in Oklahoma, which is not a seismically active area. Earthquakes, some as strong as 5.6 on the Richter scale, became more common, increasing an average of 2 a year (3.0 or greater) until 2008, to 20 in 2009, to 42 in 2010, and to 585 in 2014, triple the number in California. If all earthquakes of all magnitudes are included, Oklahoma experienced more than 5000 that year. Geologist William Ellsworth of the U.S. Geological Survey said at that time, "We can say with virtual certainty that the increased seismicity in Oklahoma has to do with recent changes in the way that oil and gas are being produced" (Galchen 2015, 35).

Nearly without exception, the stronger earthquakes are caused by fracking disposal wells, with large amounts of brackish water pumped into the ground after having been brought up during underground oil shale mining. Other states (Ohio, Arkansas, Texas, and Colorado, among others) not noted for earthquakes have also been recording more quakes in fracked areas where wastewater has been poured back into the earth, although not in the numbers that by 2015 were shaking Oklahoma with an average of two earthquakes per day.

The fear of earthquakes is real. Research published in the journal *Science* "links four of Oklahoma's most prolific wastewater wells to a swarm of ... small earthquakes near Jones, Oklahoma, by showing how the wells sent a wave of water pressure coursing through the subsurface. The pressure can reduce forces acting to keep faults locked and trigger earthquakes" (Hand 2014, 13). By 2014, Oklahoma, with its large concentration of fracking wells, was experiencing more earthquakes than California, "and seismologists are increasingly blaming them on the injection of wastewater from oil and gas operations" (Hand 2014, 13). Although "subsurface pressure data required to unequivocally link earthquakes to wastewater injection are rarely accessible," investigators writing in *Science* in 2014 linked Oklahoma earthquakes to fracking activity there (Keranen et al. 2014, 444–48).

"The Sierra Club stands with anti-fracking protesters in New Brunswick, Canada, and around the world who are protecting their land and their families from the real danger that fracking brings to the health and safety of their communities," said Sierra Club Executive Director Michael Brune. "All Canadians and all Americans should ask themselves whether a police response with tactical units and snipers was meant to serve public safety, or squelch opposition to fracking in the service of the oil industry" ("Molotov" 2013). "We are shocked by yesterday's developments and we pray for the safety of Chief Aaren Sock, his community members and other land defenders who are at the site on Elsipogtog First Nation traditional lands," said Stan Beardy, head of the Chiefs of Ontario as well as the regional chief for the Assembly of First Nations. "It is past time now to call a halt to the physical exploration work and engage Elsipogtog First Nation in a respectful dialogue. In my view, this course is in the best interests of...all concerned" ("Molotov" 2013).

7.26 Fracking Linked to Water Pollution

Protesters initiated their blockades following reports of water pollution associated with fracking. For example, in 2012, contaminants linked to fracking were detected by the U.S. Geological Survey in the groundwater of Pavillion, Wyoming, on the Wind River Indian Reservation. The contamination was detected in one of two wells drilled by the EPA in 2010. Residents of Pavillion had reported that the water had a foul smell and taste. The Indian Country Today Media Network reported that "the USGS announced that the water had again tested positive for high levels of methane, ethane, diesel compounds and phenol ... some of the chemicals used for fracking" ("Wyoming Groundwater" 2012).

Encana Corporation, a drilling company based in Calgary, asserted that the EPA had drilled into a gas reserve. The company seemed to have no explanation for the ill-smelling and rancid-tasting tap water. As of 2013, Encana had 140 natural gas wells in the area. "This goes to the heart of concerns raised by state and federal agencies, as well as Encana: EPA's wells are improperly constructed," Encana spokesman Doug Hock told the *Wall Street Journal* ("EPA Says" 2012).

The *New York Times* reported that with unemployment reaching 70% during some seasons, a debate has broken out among the Blackfeet "whether the sacredness of the mountains, streams and vegetation can be maintained if the land is drilled and its oil and gas resources extracted" ("New York Times" 2012). With at least 30 test wells being drilled, employing as many as 45–50 people each, the money is tempting. Pauline Matt told the *Times* that the drilling "threatens everything we are as Blackfeet" (Healy 2012).

"The Blackfeet are one of many tribes facing such choices," the Indian Country Today Media Network reported ("New York Times" 2012). The Turtle Mountain Band of Chippewa Tribal Council banned fracking on its reservation following a presentation to its Tribal Council by the No Fracking Way Turtle Mountain Tribe. Carol Davis, a member of the group, called its presentation "Fracking 101." "I know there's an oil boom and it's providing a lot of jobs," she added, "but we can't risk contaminating our water on the Turtle Mountain Reservation for the sake of money" ("Turtle Mountain" 2012). The Turtle Mountain band's lands, in Roulette County, North Dakota, lie atop part of the Bakken shale formation.

7.27 Oil at Fort Berthold: "The Water is Dead and It Is Lethal"

"I wanted to write a story about strength and resilience. I wanted to write a story about the singers, the horse people, and the Earth lodge builders of the Mandan, Hidatsa, and Arikara peoples, the squash and corn, the heartland of agricultural wealth in the Northern Plains," wrote Anishinaabe (Ojibway) environmental activist and author Winona LaDuke. Instead, on the northern Missouri River, she found a land remade by the Bakken oil boom. By 2014, in six years, U.S. oil production increased 70% as imports from the Organization of Petroleum Exporting Countries fell by half (Krauss 2014, B-3). Today, what remains of the Mandan, Hidatsa, and Arikara's land base makes up the Fort Berthold reservation, a grid-work of wells— about 1370 wells on the reservation, drilled and fracked, as of late 2014 (Sontag and McDonald 2014). Where "the 1954 Garrison Diversion project ... submerged a people under Lake Sakakawea, taking 152,000 acres of their best land ... drown [ing] their villages, ... agricultural wealth [and] history," the tribal council, now materially wealthy from oil and casino revenues, hosts oil executives and politicians on its 149-passenger yacht (LaDuke 2014b). The 96-foot yacht, named *Island Girl*, which cost about $2.5 million, spends most of its time out of the water, sitting on blocks, where it has become "a symbol to many of their leaders' misplaced priorities" (Sontag and McDonald 2014).

So far (as of 2014), at least 1370 wells had been drilled and fracked, at Fort Berthold. By the end of 2014, the wells were pumping more than 386,000 barrels of oil a day, one-third of North Dakota's output from the Bakken field. According to the *New York Times*, there were 850 oil-related environmental incidents at Fort Berthold reported by companies from 2007 through mid-October 2014. "Our tribal council is so focused on money, money, money," said Edmund Baker, the reservation's environmental director. "And our tribal chairman [Tex G. Hall] is: 'Edmund, don't tell me about spills. I'm busy trying to do things for my people'" (Sontag and McDonald 2014).

Tribal activists on the Fort Berthold reservation, such as Kandi Mossett, described a landscape that would be unrecognizable to their ancestors, including not only a maze of oil derricks but also "the huge Basin Electric coal generation facilities, burning the dirtiest coal in America, just upwind from their villages, [and] oil refinery proposals that have been accelerated through federal processes," along with fracking, with its appetite for precious water, as it pollutes and irradiates more than 30 trillion gallons of toxic liquid deep underground. "Simply stated," wrote LaDuke (2014b), "once water has been used in fracking, it is no longer living water. It is dead, and it is lethal." Some of this "dead" water spills onto the surface. On July 8, 2014, in what became known as the Crestwood spill, about a million gallons of saline, radioactive water leaked out of a pipe and, moving toward Lake Sakakawea, was blocked only by three beaver dams.

Despite their new wealth, many people's quality of life has deteriorated. "Apart from a significant rise in jobs, which often go to transient workers," Sontag and McDonald (2014) wrote in the *New York Times*,

> Many see deterioration rather than improvement in their standard of living. They endure intense truck traffic, degraded roads, increased crime, strained services and the pollution from spills, flares and illegal dumping.... Deep-seated problems can be hard to fix—a life expectancy of 57, for instance, compared with 79 for North Dakotans as a whole. But its critics say the tribal government has invested little in social welfare, like desperately needed housing, and has distributed little of the $200 million set aside in the People's Fund.

"Every single day more than 100 million cubic feet of natural gas is flared away," Fort Berthold tribal member and anti-fracking activist Mossett said. "That's enough to heat half a million homes. That's as much carbon dioxide [as is] emitted [by] 300,000 cars. That's crazy." The gas, which could be used as fuel, is being flared because the area lacks the required infrastructure to move it. As much as 70% of gas from wells on the reservation is flared (LaDuke 2014a).

7.28 Protests of Pipelines across Canada

Paralleling the debate over the Keystone XL Pipeline in the United States, environmentalists and representatives from more than 160 First Nations across Canada united in opposition to proposed pipelines (or expansions of existing ones) that would carry increasing production from Alberta's tar sands fields across Canada westward and eastward for export. The issues are the same as in the United States: locally, the danger of spills, and worldwide, the atmospheric pollution posed by a massive new source of carbon dioxide.

The Northern Gateway Pipeline, proposed by Enbridge, was designed to carry a daily average of roughly 525,000 barrels of diluted bitumen from the oil sands fields of Bruderheim, Alberta, to a planned marine terminal at Kitimat, British Columbia, near Vancouver, bound for export to Asia by ship. Environmentalists asserted that existing Enbridge pipelines had been responsible for 801 spills in 10 years ("Defend" 2013). The U.S. oil company Kinder Morgan also was planning to spend $5 billion, tripling capacity of the Trans Mountain Pipeline from the tar sands fields to the Vancouver, B.C. area.

TransCanada also proposed an Energy East pipeline to carry about 1.1 million barrels of crude a day from Saskatchewan and Alberta and through the Quebec City and Montreal areas to the vicinity of St. John, New Brunswick. Another pipeline would be converted from natural gas to export diluted bitumen. TransCanada announced in 2013 that it planned to spend $12 billion on these projects.

In 2013, according to the Environment News Service, "polls show[ed] 80% of British Columbians opposed crude oil tanker traffic on the BC coast," as there was an "unbroken wall of opposition" of indigenous nations, "with more than 160 First Nations putting their signatures on the Save the Fraser Declaration opposing tar sands exports projects through their territories" ("Defend" 2013). On November 16 of that year, more than 2000 people demonstrated in Vancouver against the Northern Gateway project at a "Defend Our Climate" rally.

The proposed expansion of Kinder Morgan's tar sands oil pipeline was strenuously criticized by Native peoples in Washington State and British Columbia on October 22, 2014, during hearings before Canada's National Energy Board. A pipeline terminal also was being considered at Cherry Point in northwestern Washington State, a few miles south of Vancouver. Expanding the pipeline would increase tanker loadings and put tribal fishing people at risk, "not to mention drastically increasing the chance of a catastrophic oil spill," Glen Gobin, a member of the Tulalip Board of Directors, told the National Energy Board (Hansen 2014).

A caravan of Canadian First Nations people who opposed Enbridge's Northern Gateway Pipeline, led by the Yinka Dene Alliance, formed a "Freedom Train" from Jasper, Alberta, to Toronto, that departed on April 30, 2012, to rally at Enbridge's annual meeting on May 9. The 30 native nations supporting the caravan of about 40 people said that the trip was "aimed at protecting their rights and their freedom from the threat of devastating oil spills" "Freedom Train" 2012). "Our journey on this Freedom Train is to build on the support from the people of Canada for the protection of our traditional lands. We will stand firm against all industries like Enbridge that are planning to destroy the environment and the future of all peoples," said Chief Martin Louie of the Nadleh Whut'en First Nation, a member of the Yinka Dene Alliance, in the statement. "We need a healthy environment to ensure a healthy future for our children" ("Freedom Train" 2012).

"We're traveling across Canada to tell Enbridge that they will not be permitted to build their pipelines through our lands, period," said Chief Jackie Thomas of Saik'uz First Nation. "The fight against Enbridge is a fight for our freedom to govern ourselves and to choose our own future. We will not accept the government imposing a decision on us and forcing this pipeline through our lands" ("Freedom Train" 2012).

The government of British Columbia rejected the Gateway Pipeline, having "thoroughly reviewed all of the evidence and submissions made to the panel and asked substantive questions about the project including its route, spill response capacity and financial structure to handle any incidents," said provincial Environment Minister Terry Lake. "Our questions were not satisfactorily answered during these hearings" ("Freedom Train" 2012). "The B.C. government's detailed submittal to federal government decision-makers was also noteworthy in recognizing there is growing evidence that diluted bitumen could pose additional risks to water and is more difficult to clean up," the Natural Resources Defense Council said. "In light of British Columbia's rejection, the U.S. State Department should revisit its previous findings that the Keystone XL poses no risks to water or climate which were based on faulty assumptions about the behavior of tar sands oil and about the role that Keystone XL [could play] as the major driver of tar sands strip-mining and drilling" ("Freedom Train" 2012).

In early January 2012, Canadian Prime Minister Stephen Harper accused "foreign-funded radicals" of stoking opposition to the Northern Gateway Pipeline, even as much of the opposition came from indigenous people inside of Canada. Harper said that Tides Canada, an environmental group, had received donations from the United States. "These groups threaten to hijack our regulatory system to achieve their radical ideological agenda," wrote Minister of Natural Resources Joe Oliver in an open letter a few days later. "They use funding from foreign special interest groups to undermine Canada's national economic interest" ("Ottawa Blasts" 2012).

"The minister's allegations about radicals using foreign money to achieve an ideological agenda were sweeping, and we assume he was referring to the Dene Nation and other First Nations and aboriginal organizations participating in the review process," said Dene National Chief Bill Erasmus in a statement responding to Oliver's letter. "Our mandate is to preserve and protect our communities, our land, and our culture, and it is our democratic right to participate in hearings on a pipeline that will impact us" ("Ottawa Blasts" 2012). "We have great concern about Northern Gateway and tankers coming through our traditional waters," Frank Brown, of Heiltsuk First Nation in Bella Bella, told the Indian Country Today Media Network. If necessary, we will take both legal and other means necessary to ensure we protect the traditional territories from which we come from the beginning of time" (Ball 2013a).

7.28 Protests of Pipelines across Canada

At the initial hearing on January 14, 2013, in Vancouver, more than 1000 protesters gathered outside. Inside, several speakers from Canadian First Nations and environmental groups asserted that tanker or pipeline accidents could threaten the coastline as well as nearby rivers and lands. When people made noise, the government closed the hearing to the public. Activists who snuck past police were arrested for assault or trespass, according to the Indian Country Today Media Network. "The indigenous people across whose sovereign, unceded territories this pipeline is supposed to cross have already said no, unequivocally," said Kim Heartty, who was arrested. "So the project should not be under consideration beyond that" (Ball 2013a).

"Protecting people and the environment is our top priority, which is why we announced enhancements to make what was already a very safe project even safer," said Janet Holder, an Enbridge executive vice president. "We intend to demonstrate to British Columbians and to all Canadians through the examination of the facts and science upon which this project application is based that there is a path forward that provides for prosperity while protecting the environment" (Ball 2013a).

During December 2013, Canada's federal National Energy Board and the Canadian Environmental Assessment Agency approved the Northern Gateway pipeline, setting it up for expected approval by Canada's conservative-dominated cabinet, probably to be followed by a confrontation with British Columbia's liberal provincial government as well as First Nations (whose lands may be traversed by the pipeline) and environmental groups. The review panel (which has three members) said that a major oil spill that would "have significant adverse environmental effects on ecosystems" is not likely and that "any harm . . . could be cleaned up and affected areas returned to a functioning ecosystem similar to that existing prior to the spill. . . . After weighing all the oral and written evidence, the panel found that Canada and Canadians would be better off with the Enbridge Northern Gateway project than without it" (Austen 2013).

In June 2014, the Canadian federal government's approval of the Enbridge Northern Gateway Pipeline united several hundred First Nations leaders and environmental groups in a pledge to take up legal action to stall the project. They indicated that international human rights law would be used as a litigating agent.

"The First Nations Leadership Council, which is composed of the B.C. Assembly of First Nations, First Nations Summit and Union of B.C. Indian Chiefs, is completely disgusted at this decision," said the organizations in a joint statement on June 17, 2014. "There is an undeniable and inherent risk attached to this project and the idea of a catastrophic ecological disaster is unacceptable for the people of this Province. Delaying this project will only serve to fortify the opposition to this project," said Grand Chief Stewart Phillip, president of the Union of B.C. Indian Chiefs ("Canada Approves" 2014).

A coalition said that "international human-rights standards, as set out in the *United Nations Declaration on the Rights of Indigenous Peoples* and the rulings of regional and international human rights bodies, also require a high standard of precaution in all decisions affecting Indigenous peoples' rights and their lands, territories and resources. In many instances, the standard required is that projects should proceed only with the free, prior and informed consent of Indigenous peoples" ("Canada Enbridge" 2014).

Yet another pipeline, Pacific Trails, became an object of resistance on November 27, 2012, by indigenous peoples from California and several major Canadian cities as well as in Port of Spain, Trinidad and Tobago (outside the Canadian embassy) a few days after hereditary leaders of the Wet'suwet'en Nation evicted surveyors working on the pipeline's proposed route. This pipeline was being planned to carry 1 million cubic feet of natural gas obtained through fracking 290 miles from Summit Lake, British Columbia, to Kitimat on the Pacific Coast, the same terminus as the Enbridge Northern Gateway Pipeline. Company representatives told the media that they had indigenous support along the pipeline route, but hereditary, traditional leadership had been largely ignored.

The Wet'suwet'en Nation leaders issued a traditional warning used to remove trespassers from a nation's territory. Contractors' equipment also was confiscated. "If the Pacific Trails Pipeline decides to push their agenda, along with the federal government and provincial government, to try to force this pipeline through our lands, they're going to continue to meet us, and we're going to keep resisting them," Toghestiy, a hereditary chief of Wet'suwet'en Nation, told the Indian Country Today Media Network. "If they decide to escalate it, we'll have to do the same. It's something that we don't want to do, but if they're not willing to sit down and have meaningful consultation with our hereditary chiefs, and with the Wet'suwet'en people, then they're going to be meeting a lot of resistance up here" (Ball 2012c).

Opposition to Kinder Morgan's proposed pipeline by late 2014 focused on Burnaby Mountain, a conservation area. Beginning with a camp of a few people in September, the protest grew to more than 1000 by the end of November, most of whom ignored police orders to disperse. More than 120 were arrested on civil disobedience charges when they crossed police lines. First Nations peoples joined non-Indians in the protest. On November 27, a British Columbia Supreme Court judge "refused the company's request to extend [an] injunction against disruptive protests through December 12, effectively ordering Kinder Morgan to clear its equipment out of the Burnaby Mountain Conservation Area" (Ball 2014b).

People from the Heiltsuk First Nation in Bella Bella, British Columbia, arrived, declaring "unwavering support for the keepers of the sacred fire on Burnaby Mountain.... Our relatives from the Salish Sea are demonstrating the courage and conviction of our ancestors," said Heiltsuk Chief Councilor Marilyn Slett. "Our peoples have always been keepers of the lands and waters," added Heiltsuk Hereditary Chief Harvey Humchitt. "Now we are called to be defenders as well as keepers. The land defenders on Burnaby Mountain are upholding ancestral law first and foremost, and we support them wholeheartedly" ("Tar Sands Opposition" 2014). City officials in Burnaby, a suburb of Vancouver, have sued Kinder Morgan even as the company began to test the mountain for construction in three places, drilling holes and felling trees.

The Union of British Columbia Indian Chiefs (UBCIC) also expressed opposition to the pipeline, saying that it "stands in solidarity with those that have been arrested." Union President and Grand Chief Stewart Phillip said that "we will continue to stand in support with those on the Mountain to uphold and defend Indigenous rights, land rights and human rights.... It is infuriating and beyond frustrating that we are faced with this provocative and heavy-handed approach by the RCMP when at this time the City of Burnaby's court proceedings have not even been completed. Kinder Morgan is despoiling the Burnaby Mountain Conservation Area to brazenly push ahead with their proposed expanded pipeline in the face of massive opposition" ("Tar Sands Opposition" 2014).

References

"Anishinaabe and Lakota Riders Protest Pipelines, on Horseback." 2013. Indian Country Today Media Network, October 18. http://indiancountrytodaymedianetwork.com/gallery/photo/anishinaabe-and-lakota-riders-protest-pipelines-horseback-151792. Last accessed October 23, 2013.

"Athabasca Oil Sands". 2011. NASA Earth Observatory, November 30. http://earthobservatory.nasa.gov/IOTD/view.php?id=76559&src=eoa-iotd. Last accessed December 13, 2011.

Austen, Ian. 2013. "Canadian Review Panel Approves Plans for an Oil Pipeline." New York Times, December 19. http://www.nytimes.com/2013/12/20/business/international/canadian-review-panel-approves-plans-for-an-oil-pipeline.html. Last accessed December 22, 2013.

Ball, David P. 2012a. "Athabasca Chipewyan Launch Treaty 8 Challenge to Shell Canada over Oil Sands." Indian Country Today Media Network, October 5. http://indiancountrytodaymedianetwork.com/article/athabasca-chipewyan-launch-treaty-8-challenge-to-shell-canada-over-oil-sands-137632. Last accessed October 20, 2012.

Ball, David P. 2012b. "Spiritual Leaders Vow to Defend Mother Earth from Oil Sands and Pipelines with Spiritual Declaration." Indian Country Today Media Network, November 23. http://indiancountrytodaymedianetwork.com/spiritual-leaders-vow-to-defend-mother-earth-from-oil-sands-and-pipelines-with-spiritual-declaration-147332. Last accessed December 22, 2012.

Ball, David P. 2012c. "U.S. and Canada-Wide Protests Target Pacific Trails' Proposed Fracking Pipeline." Indian Country Today Media Network, November 27. http://indiancountrytodaymedianetwork.com/article/us-and-canada-wide-protests-target-pacific-trails-proposed-fracking-pipeline-145895. Last accessed December 5, 2012.

Ball, David P. 2013a. "Enbridge Faces Rising Opposition to Northern Gateway Pipeline with Protests, Arrests." Indian Country Today Media Network, January 25. http://indiancountrytodaymedianetwork.com/2013/04/04/red-lake-band-chippewa-determined-blockade-atop-four-enbridge-pipelines-minnesota-148600. Last accessed February 5, 2013.

Ball, David P. 2013b. Fracking Troubles Atlantic First Nations after Two Dozen Protesters Arrested. Indian Country Today Media Network, June 28. http://indiancountrytodaymedianetwork.com/2013/06/28/fracking-troubles-atlantic-first-nations-after-two-dozen-protesters-arrested-150192. Last accessed July 20, 2013.

Ball, David. 2014a. "Neil Young: Blood of First Nations People Is on Canada's Hands." Indian Country Today Media Network, January 14. http://indiancountrytodaymedianetwork.com/2014/01/14/neil-young-blood-first-nations-people-canadas-hands-153104. Last accessed February 1, 2014.

Ball, David P. 2014b. "Tensions Simmer After Kinder Morgan Confrontation Over Pipeline Extension in Burnaby, B.C." Indian Country Today Media Network, December 12, 2014. Last accessed December 22, 2014. http://indiancountrytodaymedianetwork.com/2014/12/12/tensions-simmer-after-kinder-morgan-confrontation-over-pipeline-extension-burnaby-bc

Berry, Carol. 2012. Alberta Oil Sands Up Close: Gunshot Sounds, Dead Birds, a Moonscape." Indian Country Today Media Network, February 2. Last accessed February 9, 2012. http://

indiancountrytodaymedianetwork.com/article/alberta-oil-sands-up-close%3A-gunshot-sounds%2C-dead-birds%2C-a-moonscape-95444

Berry, Carol. 2013. "Tribal Members Sign Treaty Calling for an End to Alberta Oil Sands Development and Keystone XL." Indian Country Today Media Network, January 31. Last accessed February 23, 2013. http://indiancountrytodaymedianetwork.com/2013/01/31/tribal-members-sign-treaty-calling-end-alberta-oil-sands-development-and-keystone-xl

Brantley, Susan L., and Anna Meyendorff. 2013. "The Facts on Fracking." *New York Times*, March 13. http://www.nytimes.com/2013/03/14/opinion/global/the-facts-on-fracking.html. Last accessed March 16, 2013.

Call to Action! Rally at Cuomo's State of the State: No Fracking, Yes Renewables!" 2014. Two Row Wampum Renewal Campaign, January 14. http://honorthetworow.org/call-to-action-rally-at-cuomos-state-of-the-state-no-fracking-yes-renewables. Last accessed July 3, 2014

"Canada Approves Enbridge Pipeline Through B.C., First Nations Will Sue". 2014. Indian Country Today Media Network, June 19. http://indiancountrytodaymedianetwork.com/2014/06/19/canada-approves-enbridge-pipeline-through-bc-first-nations-will-sue-155377

"Canada Enbridge Approval Unites First Nations, Amnesty, Rights Groups in Opposition". 2014. Indian Country Today Media Network, June 20, 2014, p. A-1.

Capriccioso, Rob. "Indigenous Oil Sands Protest Leads to White House Arrests." Indian Country Today Media Network. September 2, 2011. A-4

"Chiefs Declare Keystone XL Consultation Meeting Invalid, Walk Out on State Department Officials". 2013. Indian Country Today Media Network, May 17, 2013.

Dadigan, Marc. 2013. "Umatilla Tribe Battles Mega-Loads Headed for Alberta Oil Sands." Indian Country Today Media Network, December 1. 2013, p. A-2.

"'Defend Our Climat eRallies Draw Thousands across Canada". 2013. Environment News Service, November 18, 2014.

"Elsipogtog Chief Issues Eviction Notice to Texas-Based Frackers. Band Council Resolution to Reclaim All Unoccupied Crown Land". August 5, 2013. Halifax Media Co-op, October 4, 2013, A-7.

"Emissions from Oil Sands Mining". 2012. NASA Earth Observatory, March 2. http://earthobservatory.nasa.gov/IOTD/view.php?id=77283&src=eoa-iotd. Last accessed March 20, 2012.

"EPA Says Wyoming Fracking Results Are Consistent". 2012. *Wall Street Journal,* September 26. . Last accessed October 11, 2012. http://online.wsj.com/article/SB10000872396390443328404578020923049282436.html

Erlanger, Steven. 2013. "As Drilling Practice Takes Off in U.S., Europe Proves Hesitant." *New York Times*, October 9, 2013,

"Every Tree and Plant Died': Massive Toxic Spill Guts Alberta". 2013. Indian Country Today Media Network, August 26. http://indiancountrytodaymedianetwork.com/2013/08/26/toxic-wastewater-spill-alberta-kills-dene-tha-landscape-150968. Last accessed February 22, 2015.

"Fill Gaps in Keystone XL Draft Environment Report or Reject Pipeline, NCAI Tells Obama Administration". 2013. Indian Country Today Media Network, May 2, 2013, A-3.

"'Freedom Train' Protesting Northern Gateway Pipeline Crossing Canada." 2012. Indian Country Today Media Network, April 30. http://indiancountrytodaymedianetwork.com/most-popular?page=2677. Last accessed May 5, 2012.

Frosch, Dan. 2013. "Amid Pipeline Debate, Two Costly Cleanups Forever Change Towns." *New York Times,* August 10. http://www.nytimes.com/2013/08/11/us/amid-pipeline-debate-two-costly-cleanups-forever-change-towns.html. Last accessed August 14, 2013.

"Full Text of the Great Plains Tribal Chairmen's Association Statement against the Keystone XL Pipeline". 2013. Indian Country Today Media Network, May 17. http://indiancountrytodaymedianetwork.com/2013/05/17/full-text-great-plains-tribal-chairmens-association-statement-against-keystone. Last accessed May 23, 2013.

Galchen, Rivka. 2015. "Weather Underground." *The New Yorker*, April 13, 34–40.

References

Hand, Eric. 2014. "Injection Wells Blamed in Oklahoma Earthquakes." *Science* 345 (July 4): 13–14.

Hansen, James. 2012. "Game Over for the Climate." *New York Times*, May 9. http://www.nytimes.com/2012/05/10/opinion/game-over-for-the-climate.html. Last accessed May 15, 2012.

Hansen, Terri. 2014. "Coast Salish Unite against Tripling Capacity of Kinder Morgan Tar Sands Pipeline." Indian Country Today Media Network, October 27. Indiancountrytodaymedianetwork.com/2014/10/27/coast-salish-unite-against-tripling-capacity-kinder-morgan-tar-sands-pipeline-157543. Last accessed November 4, 2014.

Healy, Jack. 2012. "Tapping into the Land and Dividing Its People." *New York Times*, August 15. http://www.nytimes.com/2012/08/16/us/montana-tribe-divided-on-tapping-oil-rich-land.html. Last accessed August 22, 2012.

Homer-Dixon, Thomas. 2013. "The Tar Sands Disaster." *New York Times*, March 31. http://www.nytimes.com/2013/04/01/opinion/the-tar-sands-disaster.html. Last accessed April 11, 2013.

Howe, Miles. 2013. "RCMP Bring 60 Drawn Guns, Dogs, Assault Rifles, to Serve Injunction on the Wrong Road." Halifax Media Co-op, October 18.

Johansen, Bruce E. "Dakota Conflict the Latest of Many." *Omaha World-Herald*, December 13, 2016. http://www.omaha.com/opinion/bruce-e-johansen-dakota-conflict-the-latest-of-many/article_4aaf6af6-3892-5d04-bf60-4c2ee49e46a5.html. Last accessed May 12, 2022.

Johnson, Kirk. 2013. "Fight over Energy Finds a New Front in a Corner of Idaho." *New York Times*, September 25. http://www.nytimes.com/2013/09/26/us/fight-over-energy-finds-a-new-front-in-a-corner-of-idaho.html. Last accessed September 27, 2013.

Keranen, K. M., M. Weingarten, G. A. Abers, B. A. Berkins, and S. Ge. 2014. "Sharp Increase in Central Oklahoma Seismicity since 2008 Induced by Massive Wastewater Injection." *Science* 345 (July 25): 444–48.

Klein, Naomi. 2014. *This Changes Everything: Capitalism and the Climate*. New York: Simon & Schuster.

Koch, Wendy. 2014. "Would Keystone Pipeline Unload 'Carbon Bomb' or Job Boom?" *USA Today*, March 10. http://www.usatoday.com/story/news/nation/2014/03/01/keystonexls-myths-debunked/5651099. Last accessed March 24, 2014.

Kolbert, Elizabeth. 2007. "Unconventional Crude: Canada's Synthetic-Fuels Boom." *The New Yorker*, November 12, 46–51.

Krauss, Clifford. 2013. "Looking for a Way around Keystone XL, Canadian Oil Hits the Rails." *New York Times*, October 31. http://www.nytimes.com/2013/10/31/business/energy-environment/looking-for-a-way-around-keystone-xl-canadian-oil-hits-the-rails.html. Last accessed November 4, 2013.

Krauss, Clifford. 2014. "U.S. Oil Prices Fall below $80 a Barrel." *New York Times*, November 4, B-3.

Krauss, Clifford, and Ian Austen. 2014. "Rocky Road for Canadian Oil." *New York Times*, May 13, B-1, B-4.

Krauss, Clifford, and Jad Mouawad. 2014. "Accidents Surge as Oil Industry Takes the Train." *New York Times*, January 26, A-1.

"Lac-Mégantic Rail Tragedy Resonates at Quinault Nation". 2013. Indian Country Today Media Network, July 29, 2013, p. 6-A.

LaDuke, Winona. 2014a. "Bakken Gas Flares Away, as Nationwide Propane Shortage Kills with Hypothermia." Indian Country Today Media Network, October 9. http://indiancountrytodaymedianetwork.com/2014/10/09/bakken-gas-flares-away-nationwide-propane-shortage-kills-hypothermia-157249. Last accessed October 15, 2014.

LaDuke, Winona. 2014b. "'Unspeakable Poverty of Loss': Intergenerational Trauma and the Bakken Oil Fields." Indian Country Today Media Network, October 8. http://indiancountrytodaymedianetwork.com/2014/10/08/unspeakable-poverty-loss-intergenerational-trauma-and-bakken-oil-fields-157243. Last accessed October 13, 2014.

"Leo DiCaprio Tours Tar Sands, Joins Natives in ALS Ice Bucket Challenge". 2014. Indian Country Today Media Network, August 26. http://indiancountrytodaymedianetwork.

com/2014/08/26/leo-dicaprio-tours-tar-sands-joins-natives-als-ice-bucket-challenge-156602. Last accessed August 31, 2014.

Leslie, Jacques. 2014. "Is Canada Tarring Itself?" *New York Times*, March 31, A-19.

Lizza, Ryan. 2013. "The President and the Pipeline." *The New Yorker*, September 16, 38–51.

"Lower Brule Sioux Reject Keystone XL, Evict TransCanada". 2015. Indian Country Today Media Network, May 1. http://indiancountrytodaymedianetwork.com/2015/05/01/lower-brule-siouxreject-keystone-xl-evict-transcanada-160214. Last accessed May 12, 2015.

McNeel, Jack. 2015. "Too Many Broken Pipelines: Fort Peck Reservation Passes Resolution Opposing Keystone XL through Montana." Indian Country Today Media Network, February 24. Last accessed March 1, 2015. Indiancountrytodaymedianetwork.com/2015/02/24/toomany-broken-pipelines-fort-peck-reservation-passes-resolution-opposing-keystone-xl

"Message in a Blanket: Dakota Artist Stealth-Embroiders 'NOKXL' into Obama Quilt". 2015. Indian Country Today Media Network, May 16.

"Molotov Cocktails and Guns Confiscated; Support Pours in for Mi'kmaq Protesters". 2013. Indian Country Today Media Network, October 18.

Morin, Brandi. 2013. "Federal and Provincial Governments Lose Appeal against First Nation Oil Sands Lawsuit." Indian Country Today Media Network, May 8. http://indiancountrytodaymedianetwork.com/2013/05/08/federal-and-provincial-governments-loseappeal-against-first-nation-oil-sands-lawsuit. Last accessed July 3, 2013.

"NASA Scientist, Religious Leaders Arrested in Tar Sands Protest". 2011. Environment News Service, August 29. http://www.ens-newswire.com/ens/aug2011/2011-08-29-02.html. Last accessed September 2, 2011.

"NCAI Condemns Keystone XL Pipeline". 2011. Indian Country Today Media Network, August 18.http://indiancountrytodaymedianetwork.com/article/ncai-condemns-keystone-xl-pipeline-4 8112. Last accessed August 24, 2011.

The New York Times Profiles Blackfeet Drilling Dilemma. 2012. Indian Country Today Media Network, August 16. http://indiancountrytodaymedianetwork.com/article/%3Cem%3Ethenew-york-times%3C/em%3E-profiles-blackfeet-drilling-dilemma-129595. Last accessed August 20, 2012.

"Nez Perce Victory: U.S. Forest Service Forbids Mega-Loads along Highway 12". 2013. Indian Country Today Media Network, September 20.

Nikiforuk, Andrew. 2014. "Large Dams of Mining Waste Leaking into Athabasca River: Study." Indian Country Today Media Network, February 26. indiancountrytodaymedianetwork.com/2014/02/26/large-dams-mining-waste-leaking-athabasca-river-study-153749?page=0%2 C2. Last accessed March 3, 2014.

"Ottawa Blasts 'Radicals,' Blames 'Foreign Interests' for Northern Gateway Opposition". 2012. Indian Country Today Media Network, March 14. http://indiancountrytodaymedianetwork.com/article/ottawa-blasts-%27radicals%2C%27-blames-%27foreign-interests%27-for-north ern-gateway-opposition. Last accessed March 20, 2012.

Pember, Mary Annette. "Enbridge Takes the Gloves off In Line 5 Battle." *Indian Country Today*, May 8, 2022, 1. Last accessed May 28, 2022.

Schilling, Vincent. 2012. "Pine Ridge Residents Halt Canadian Mine Equipment Transportation through Reservation." Indian Country Today Media Network, March 9. http://indiancountrytodaymedianetwork.com/article/pine-ridge-residents-halt-canadian-mine-equipment-transportation-through-reservation-102246. Last accessed March 13, 2012.

Schilling, Vincent. 2013a. "Behind the Front Lines of the Elsipogtog Battle over Fracking." Indian Country Media Network, November 20. http://indiancountrytodaymedianetwork.com/2013/11/20/behind-lines-elsipogtog-aptn-journalist-ossie-michelin-gives-backstory-1523 54. Last accessed November 29, 2013.

Schilling, Vincent. 2013b. "Woman with Eagle Feather: The Photo 'Heard' Round the World." Indian Country Today Media Network, November 21. http://indiancountrytodaymedianetwork.com/2013/11/21/woman-eagle-feather-photo-heard-round-world-152357. Last accessed November 26, 2013.

References

Smith, Mitch. 2015. "Grass-roots Push in the Plains to Block the Keystone Pipeline's Path." *New York Times*, May 6. http://www.nytimes.com/2015/05/06/us/grass-roots-push-in-the-plains-to-block-the-keystone-pipelines-path.html. Last accessed May 10, 2015.

Sontag, Deborah, and Brent McDonald. 2014. "In North Dakota, a Tale of Oil, Corruption and Death." *New York Times*, December 29. Last accessed January 5, 2015.

Stone, Laura. 2013. "Chief Tale Slams RCMP for 'Extreme Use of State Force' in N.B. Shale Gas Protest." Global News (Canada), October 18. http://globalnews.ca/news/910934/chief-atleo-slams-rcmp-for-extreme-use-of-state-force-in-n-b-shale-gas-protest/?utm_source=facebook-twitter&utm_medium=link&utm_campaign=community. Last accessed October 25, 2013.

"Tar Sands Opposition Unites Burnaby Leaders, First Nations, as 60 Arrested in B.C. Protests". 2014. Indian Country Today Media Network, November 26. http://indiancountrytodaymedianetwork.com/2014/11/26/tar-sands-opposition-unites-burnaby-leaders-first-nations-60-arrested-bc-protests-158031. Last accessed December 3, 2014.

Taylor, Kevin. 2013. "Nez Perce Leaders Stand Firm on Frontlines of Mega-Load Transport." Indian Country Today Media Network, August 9. Last accessed August 15, 2013. http://indiancountrytodaymedianetwork.com/2013/08/09/nez-perce-leaders-stand-firm-frontlines-mega-load-transport-15080

Tollefson, Jeff. 2013. "Climate Science: A Line in the Sands." *Nature* 500 (August 8): 136–37. http://www.nature.com/news/climate-science-a-line-in-the-sands-1.135150. Last accessed August 26, 2013.

Troian, Martha. 2013a. "First Nation Moves to Evict Fracking Co. from Lands Held in Trust." Indian Country Today Media Network, October 3. http://indiancountrytodaymedianetwork.com/2013/10/03/elsipogtog-first-nation-wrests-lands-fracking-co-mikmaq-treaty-day-151564. Last accessed October 11, 2013.

Troian, Martha. 2013b. "Police in Riot Gear Tear-Gas and Shoot Mi'kmaq Protesting Gas Exploration in New Brunswick." Indian Country Today Media Network, October 17. http://indiancountrytodaymedianetwork.com/2013/10/17/police-riot-gear-tear-gas-and-shoot-mikmaq-protesting-gas-exploration-new-brunswick. Last accessed October 30, 2013.

Troian, Martha. 2014. "Court Tells Lubicon Lake Cree in Alberta to Stop Blocking Fracking Activities." Indian Country Today Media Network, February 6. http://indiancountrytodaymedianetwork.com/2014/02/06/court-tells-lubicon-lake-cree-alberta-stop-blocking-fracking-activities-153435. Last accessed February 12, 2014.

"Turtle Mountain Tribal Council Bans Fracking". 2012. Indian Country Today Media Network, November 27. http://indiancountrytodaymedianetwork.com/article/turtle-mountain-tribal-council-bans-fracking-64866. Last accessed December 12, 2012.

Woodard, Stephanie. 2013. "Planned Oil Pipeline Must Cross Pine Ridge's Water-Delivery System." Indian Country Today Media Network, September 21. http://indiancountrytodaymedianetwork.com/article/planned-oil-pipeline-must-cross-pine-ridge%2 5e2%2580%2599s-water-delivery-system-54942.html. Last accessed September 30, 2013.

"Wyoming Groundwater Again Tests Positive for Fracking-Related Chemicals on Wind River Reservation". 2012. Indian Country Today Media Network, September 27. Last accessed October 3, 2012. http://indiancountrytodaymedianetwork.com/article/wyoming-groundwater-again-tests-positive-for-fracking-related-chemicals-on-wind-river-reservation-136259

Chapter 8
Mining: Tearing at Mother's Breast

To many Native Americans and their non-Native allies—those with a naturalistic philosophy—mining is the ultimate insult, a rape of mother earth transformed into mother lode. Thus, the number and intensity of protests by indigenous peoples against mining across Turtle Island (North America). The mining of uranium is such a natural breach that Navajo cosmology warns against it.

Mercantile capitalism has no such qualms about mining. It is, in fact, the very basis of a system that survives and thrives by making and selling things. Thus, the germ of conflict: Native American homelands contain vast stores of exploitable natural resources that corporations find useful and profitable, often available from governments at prices far below market value. Or so they were—until the "other" started fighting back.

8.1 The World's Largest Salmon Run Vs. the World's Biggest Gold and Copper Mine

The catalog of mining and its toxic legacy on Native American lands is vast. Many Native peoples across North America have organized against mining on their homelands, notably in Alaska, where the world's largest sockeye salmon run could be imperiled by the proposed Pebble Mine, as the largest gold and copper strip mine on Earth, which has the potential to produce an incredible amount of gold and copper, as well as much more rock "overburden."

The Pebble Partnership proposed a 1700-foot-deep, two-mile-wide open-pit mine (with underground mines as well) 200 miles southwest of Anchorage, Alaska, that would dump mining wastes in some of the world's most fertile salmon spawning grounds, home of the Earth's largest surviving wild sockeye salmon run. Twenty-five federally recognized Alaskan Native tribes live in the Bristol Bay watershed, having maintained a culture and economy based on salmon for at least 4000 years.

On average, 37 million sockeye return each year to Bristol Bay, supporting more than 12,000 Eskimo and non-Native commercial fishing people. The Pebble Mine could turn part of this salmon run's natural end into a toxic dump.

Joby Warrick of the *Washington Post* (2015) described "The region's marshy lowlands [that] are dotted with kettle lakes and crisscrossed by countless streams. In the spawning season, icy rivers turn into roiling torrents of red fish as tens of millions of Pacific salmon make their way inland to spawn." The area has been known as a rich potential source of gold for 30 years, but the day is past when freelancers could pan chunks of gold from streams or dig them out of the earth. The Pebble Mine's riches are embedded in fine particles throughout the soil.

"To extract the gold profitably," wrote Warrick, requires excavating huge amounts of soil and rock from an open-pit mine that could, according to preliminary design plans, eventually cover a seven-square-mile area, making it one of the world's largest. The project would create mountains of rocky spoils, while the wastewater from extracting the ore would be kept in large containment ponds. This is true of other strip-mined minerals as well. The more intensely they are mined, the smaller the percentage of yield, and the greater the waste rock left over.

To the Taseko mining company, a massive strip mine for gold and copper meant "new prosperity," but to the Tsilhqot'in (Chilcotin) indigenous people of British Columbia it meant genocide. They united with environmentalists and other indigenous peoples across Canada and won a reprieve. "As a Nation, we've made a decision to focus our energies on the real battle of defeating this project, full-stop," said Chief Marilyn Baptiste of the Xeni Gwet'in, one of the communities comprising the Tsilhqot'in National Government (TNG).

Over several years, the Tsilhqot'in and their allies forced the scaling down (and finally cancelation, at least for a time) of the mine. An early proposal would have severely damaged the trout fishery In *Teztan Biny* (Fish Lake). The new plan said it would spare that area, but still would have turned a smaller body of water, Little Fish Lake, into a tailings dump. To stop it, opponents went to court and established a blockade. "It's a simple *No*," said Marilyn Baptiste, chief of Xeni Gwet'in, the Tsilhqot'in band nearest the proposed mine site. "Our land is not for sale. Our position has not changed and cannot change for the destruction of our lands, our waters, and our way of life. Our wild rainbow trout has survived in that lake system for hundreds of years, as our people have. We will not, and cannot, agree to such destruction in the headwaters by Taseko".

The company asserted that its plans were environmentally responsible, but the opponents did not relent. In June 2012, Taseko delivered a revised proposal at its annual meeting in Vancouver, B.C., which mitigates environmental damage, according to President and Chief Executive Officer Russell Hallbauer. "This project ... holds exciting potential for the company's stakeholders, including shareholders and local communities," Hallbauer wrote. "These revised plans address the environmental concerns identified in the original environmental assessment process, and importantly, includes the preservation of Fish Lake".

In November 2020, the U.S. Army Corps of Engineers rejected Northern Dynasty Minerals' permit application to dig Pebble Mine; A permit for that mine also had been denied in 2014 by the U.S. Environmental Protection Agency (EPA), but mining companies have challenged both decisions, and no doubt will not relinquish their pursuit of a gigantic payoff. Resistance from Alaskans and Native peoples will continue as well. The United States' federal government's position on Pebble Mine has switched with whomever holds power: Out under Obama, in with Trump, out with Joe Biden, as of 2022.

Conflicts over mining and Native Americans' role in it spans North America from sea to shining sea. What follows is a brief summary. In Wisconsin, copper, zinc, and sulfide mining has been proposed and resisted by Native peoples. Silver mining in Mexico has been poisoning indigenous children. Coal strip mining has been an important issue for Hopis and Navajos, where many shepherds have lost many of their sheep to the toxic plumes of the Four Corners power plants, a substantial number of which have now been closed (as of 2022) because of pressure from Navajos and non-Indian environmentalists, as well as changing energy markets.

Anglo-American, a mining conglomerate headquartered in London, proposed to join with British Columbia's Northern Dynasty Minerals in digging the mine near Bristol Bay, Alaska, with facilities to crush and separate rock, as well as tailings (waste) dumps that would eventually become far larger than the mine itself, which would become the largest open-pit mine in North America. Potential yield over the life of the mine has been estimated at 110 million ounces of gold and 80 billion pounds of copper (Bristol Bay 2012). Even so, the mining companies insisted that salmon runs and wildlife habitat would not be harmed. An Environmental Protection Agency assessment issued in March 2012 came to a different conclusion: "The project would entail the destruction of 55 to 87 miles worth of thus-far-untouched streams and 2,500 acres of wetlands—and that doesn't even begin to address the potential for disaster if any of the tailings ponds were to leak their acidic water and heavy metals into salmon spawning grounds" (Bristol Bay 2012). The EPA assessment was a best case, assuming that earthen dams (some of them 700 feet in height) that hold tailings ponds intact would not break or leak.

Callan J. Chythlook-Sifsof, a Yupik/Inupiat Eskimo who was raised in Aleknagik, Alaska, an indigenous Yupik Eskimo Village near the site of the proposed mine, recalled his youth in a community whose members fished for a living, before he became a member of the U.S. snowboarding team in the 2010 Winter Olympics.

> I spent my summers on the back deck of family fishing boats working multiple fisheries. The boats and fish camps are maintained by generations of families harvesting salmon not only for income, but also for food. I remember long days of processing hundreds of pounds of salmon, setting nets, cleaning and filleting, filling tubs of salt brine, putting fresh water in clean white buckets and hanging neat rows to dry and smoke.... As a child, I had no idea what magic this life was—it was just the way we did things. It's the way many

(continued)

> Alaska Natives live—through self-reliance and hard work to harvest the many gifts of the land and sea.
> This subsistence way of life that is thousands of years old is threatened by the plans of a British and Canadian mining partnership to dig a huge mine in the heart of our productive, healthy watershed. Risking our lands, waters, and fish for a short-term mega-mine like Pebble is a terrible idea. Eighty-one per cent of Native shareholders in the Bristol Bay Native Corporation—composed of more than 9000 Native Alaskans with ancestral ties to the Bristol Bay region—opposed the mine in a 2011 survey. And that's despite the promises of jobs and continuing efforts by the Pebble Partnership...to buy support through grants and giveaways to communities and hundreds of millions spent to develop the mine (Chythlook-Sifsof 2013).

In a survey by the Natural Resources Defense Council, 68% of Alaskans opposed the mine. More than 75% of 900,000 public comments on the project collected by the U.S. Environmental Protection Agency opposed the project. "It's truly alarming," wrote Chythlook-Sifsof, "when Pebble's chief executive officer, John Shively, blithely says that, sure, the mine will damage some salmon habitat, but the company will just build 'comparable' habitat nearby. His comments show a massive lack of understanding about salmon life cycles, habitat and ecosystems—not to mention the people of Bristol Bay" (Chythlook-Sifsof 2013). Native peoples urged the E.P.A. to reject plans for the mine under the aegis of the federal Clean Water Act.

A total of 360 scientists from the United States, Poland, Canada, Australia, and other countries on February 3, 2014 signed a letter urging the E.P.A. to disapprove the Pebble Mine development. "Over three years, EPA compiled the best, most current science on the Bristol Bay watershed to describe how large-scale mining could impact salmon and water in this unique area of unparalleled natural resources," said Dennis McLerran, regional administrator for E.P.A. Region 10.

"Our report concludes that large-scale mining poses risks to salmon and the tribal communities that have depended on them for thousands of years. Based on the results of the assessment, we are very concerned about the prospect of large-scale mining in the unique and biologically rich watersheds of southwest Alaska's Bristol Bay," the scientists wrote. "The preponderance of evidence presented in the watershed assessment indicates that large-scale hard rock mining in the Bristol Bay watershed threatens a world-class fishery and uniquely rich ecosystem, and we urge the Administration to act quickly to protect the area. Therefore, we urge [the] EPA to use its authority under the Clean Water Act to take the necessary next steps to protect Bristol Bay" (360 Scientists 2014).

Anglo-American in September 2013 withdrew from an industrial consortium seeking to develop the Pebble Mine. Anglo-American had been half owner with a mineral exploration and development company, Northern Dynasty, based in Vancouver, B.C. "Our focus has been to prioritize capital to projects with the highest value and lowest risks within our portfolio, and reduce the capital required to sustain such projects during the pre-approval phases of development as part of a more effective, value-driven capital allocation model," Mark Cutifani, chief executive of Anglo-American (Anglo American Withdraws 2013). In other words, the battle between the company and its opponents had become too expensive for Anglo-American to continue profitably.

Following the appeal by the 360 scientists, as well as a visit to Washington, D.C. by representatives of the 31 Native villages in the Bristol Bay watershed—all adamantly opposed to the Pebble mine proposal—the EPA late in February 2014 refused to allow the Army Corps of Engineers to issue a permit to Northern Dynasty Minerals Ltd. of Canada. "Extensive scientific study has given us ample reason to believe that the Pebble Mine would likely have significant and irreversible negative impacts on the Bristol Bay watershed and its abundant salmon fisheries," EPA Administrator Gina McCarthy said. "It's why [the] EPA is taking this step forward in our effort to ensure protection for the world's most productive salmon fishery from the risks it faces from what could be one of the largest open pit mines on earth. This process is not something the Agency does very often, but Bristol Bay is an extraordinary and unique resource." The EPA statement pointed out that water quality is crucial to the area's average annual run of 37.5 million Sockeye salmon, as well as smaller runs of Chinook, coho, pink, and chum salmon. "In addition, it is home to more than 20 other fish species, 190 bird species, and more than 40 terrestrial mammal species, including bears, moose, and caribou," the EPA said. Digging an open-pit gold and copper mine a mile deep and 2.5 miles wide would require at least three earthen tailings dams as high as 650 feet that "would irreparably harm the ecosystem" (Native Alaskans 2014).

Before dawn August 4, 2014, three weeks after the Pebble Mine's initial rejection, as its sponsors were getting ready to resubmit the proposal, a tailings pond burst in British Columbia's Mount Polley Mine, sending 4 billion gallons of mining waste (2.6 billion gallons of water mixed into a slurry with 1.2 billion gallons of fine sediment laced with metals) into rivers and lakes awaiting the return of salmon. Knight Piesold Consulting, the firm that had designed the burst dam had also been hired as a contractor to design the tailings pond for the Pebble Mine, according to a report filed in 2006 by Northern Dynasty Mines Inc. with the Alaska Department of Natural Resources.

Critics of the Pebble Mine proposal saw the Mount Polley Mine tailings spill as a prelude and a warning. "We don't want this to happen in Bristol Bay," said Kim Williams, of Dillingham, director of Nunamta Alukestai, which means Caretakers of Our Land, a conservation group of Alaska Native tribes and corporations, to *The Cordova Times*. "With all the similarities between Pebble and the Mount Polley copper mine, we're urging the EPA to take immediate action to finalize mine waste restrictions in Bristol Bay. Our hearts go out to those in British Columbia who live downstream from this devastating mine failure....It's Bristol Bay's worst nightmare," said Carol Ann Woody, a Center for Science in Public Participation fisheries scientist (Company 2014).

"The Pebble project would be bigger—a lot bigger," wrote Joel Reynolds, western director and senior attorney for the National Resources Defense Council in the *Huffington Post*. "While Imperial Metals has been mining about 20,000 tons per day at its Mount Polley mine, Northern Dynasty has anticipated about ten times that much at Pebble, with a tailings pond many times larger in footprint and scale" (Company 2014).

After Imperial Metals President Brian Kynoch bragged that he would drink the water downstream after the Mount Polley spill, the Indian Country Today Media Network headlined: "Pour him a Tall One!" (Pour 2014). "It's very close to drinking water quality, the water in our tailings," Kynoch said, according to the Canadian Broadcasting Corporation (CBC) "There's almost everything in it but at low levels.... No mercury [and] very low arsenic."

Several First Nations people live in the area of the spill and depend on its salmon runs. "Our communities are filled with sorrow, frustration and anger as they are left wondering just what poisons are in the water, and what is being done to address this disaster," said Williams Lake Chief Ann Louie and Xat'sull Chief Bev Sellars. "Monday's devastating tailings pond breach is something that both [of] our First Nations have lived in fear of for many years," the chiefs said. "We have raised repeated concerns about the safety and security of this mine, but they were ignored. Now we are being ignored again. Enough is enough" (Horrific 2014).

NASA's Earth Observatory described the Polley Mine dam breach as viewed from satellites high above the Earth:

> On August 5, nearly all of the wastewater in the retention basin had drained, exposing the silty bottom. Hazeltine Creek, usually about 1 meter (3.3 feet) wide, swelled to a width of 150 meters (490 feet) as a result of the spill. In the aftermath of the flood, a layer of brown sediment coated forests and stream valleys affected by the spill. Notice how much forest immediately north of the retention basin was leveled. Debris, mainly downed trees, are visible floating on Quesnel Lake....The breach released more than 10 million cubic meters (350 million cubic feet) of water and 4.5 million cubic meters (150 million cubic feet) of sand into Polley and Quesnel Lake, according to the British Columbia's Ministry of Environment. That is enough water to fill 4000 Olympic-sized pools (Dam Breach 2014).

Native peoples were not persuaded by Imperial Metals' insistence that the mine waste behind their ruined dam was harmless. "As of last night, Department of Fisheries and Oceans has banned salmon fishing in the Cariboo and Quesnel Rivers due to Mount Polley," said Chief Bob Chamberlin, vice president of the Union of B.C. Indian Chiefs, said on August 6. "Mount Polley will have an immediate and devastating effect on First Nations like Lhtako Dene, and Lhoosk'uz Dene. Nazkoand Esdilagh [and other Native fishing peoples] may not be able to fish for salmon at all this year. First Nations are anxiously awaiting the water-test results, the possible... closures afterwards and the harmful impacts on future salmon runs of the Fraser" (Horrific 2014).

Under the Obama administration, the E.P.A. had committed three years to study of the mine proposal, gathering several million public comments, most of them opposed. Alaskans, living in a state where a sustainable fishery is written into the state constitution, were overwhelmingly opposed, in support of the commercial fishing people and Native Americans who have lived and fished in the area for 10,000 years. Someone added up the number of salmon that the Bristol Bay fishery had provided to Alaskans and came up with an estimate of 2 billion. "It took 95 years to catch the first billion, and just 38 years to catch the second," wrote Brendan Jones in the *New York Times* (2017). Alaska U.S. Senator Ted Stevens, a Republican who usually sided with miners, opposed the project. President Obama's E.P.A. agreed, concluding that Pebble Mine would commit "irreversible" damage to the wetlands, resulting in "complete loss of fish habitat due to elimination, dewatering and fragmentation of streams, wetlands and other aquatic resources" (Jones 2017).

Then, in 2016, Donald Trump was elected president and appointed Scott Pruitt to head the E.P.A. Jones, a commercial fisherman at Bristol Bay wrote in the *New York Times* that, "On May 1 [2017], Scott Pruitt, the new administrator of the E.P.A., met with Tom Collier, the head of the Pebble Mine project [of Northern Dynasty Minerals], for breakfast. Later that same morning, Pruitt ordered the E.P.A. regulations scrapped, telling the company it could proceed with permitting. And like that, the mine was back in play" (Jones 2017).

Northern Dynasty originally had proposed to blow the largest manufactured hole on Earth in one of the planet's last large salmon fisheries, and call it "sustainable." The revised proposal called for a smaller hole but, as Jones noted, "Once drilling begins, permits are easily amended. What mine in the history of the world has ever left gold in the ground? The company also proposes that the mine's toxic wastewater be kept in a reservoir protected by an earthen dam—in one of North America's most active earthquake zones" (Jones 2017).

8.2 Common Cause at Standing Rock

Standing Rock (Chap. 7) was the best known of many recent pipeline protests that pitted Native peoples against oil and gas interests. They have been acts of cathartic, communal expression, and shared outrage against the defilement of Mother Earth by a way of thinking that regards nature as a vending machine: mother lode versus Mother Earth. Judith LeBranc, a member of the Caddo Nation of Oklahoma and director of the Native Organizers Alliance, which trains social activists in Indian Country, described the sense of common cause shared by protesters:

> "At Standing Rock, our love and commitment to the land, to our people, and to all the people of Mother Earth, made us much stronger when we were facing

(continued)

> rubber bullets, when we were being hosed down [in freezing weather], and watching drones flying overhead. It's harder in some ways to resist brutality with love, but it builds a core strength that can never be defeated. That's the kind of moral resistance that people are yearning for. They are yearning for unity and a sense of humanity that will give us the strength necessary to survive very difficult conditions. If we're going to change the system, if we're going to have economic and racial justice, it's going to take resistance driven by compassion....Coming together showed the world we were willing to do whatever it took to protect Mother Earth and the sacred, not just for ourselves, but for everyone on the planet....At the high point [at Standing Rock], 10,000 of us were bound together by a common belief..." (LeBlanc 2018, 213–214).

Aside from the delicious historical irony of a "cowboy-Indian" alliance, common cause among Native Americans and environmentally aware European-Americans (especially against such projects as the Keystone XL and Dakota Access pipelines) is quite logical. Both groups have common interests in maintaining a sustainable home. The months of solidarity between diverse peoples as they were sprayed with cold water in freezing weather illustrates just how strong these human bonds with Mother Earth became. The "cowboy/Indian" alliance tied TransCanada into political and legal knots in Nebraska over the Keystone XL, as described by Zoltán Grossman in *Unlikely Alliances: Native Nations and Communities Join to Defend Rural Lands*. Seattle: University of Washington Press, 2017.

Grossman, who teaches geography and Native American Studies, at the Evergreen State University (near Olympia, Washington) wonderfully describes the special bond that formed between Native peoples and white ranchers against the Keystone pipeline in Nebraska. The author misses an important part in the history of interethnic alliances on vital issues that occurred in the same city (Seattle) where the book was published. While the author acknowledges that these Native/non-Native environmental alliances … began in the Pacific Northwest in the 1970s, Native American, black, Latino, Asian, and progressive white allies supported each other in and near Seattle, founding important institutions such as the Daybreak Star Center and El Centro de la Raza (among others). Natives who were fishing at considerable personal risk before the Boldt decision (1974) offered hundreds of pounds of smoked salmon to feed people who occupied land and buildings in and near Seattle to create urban infrastructure, such as El Centro de la Raza.

Protests also built again in 2018 throughout the Pacific Northwest against expansion of a 700-mile-long Kinder Morgan's Trans Mountain pipeline project that was being planned to carry 890,000 barrels a day of Alberta tar sands to an export terminal in Burnaby, near Vancouver, B.C, then to ports around the Pacific Rim. The new pipeline would have tripled the existing capacity of the Trans Mountain Pipeline along the same route.

8.2 Common Cause at Standing Rock

On March 10, 2018, about 5000 people took to the streets in Vancouver to protest major oil-pipeline expansion in Canada and the United States. "We cannot stand by anymore," said Will George of the Tsleil-Waututh Nation, spokesman for the Protect the Inlet movement. "It is going to be like Standing Rock," where several thousand of water protectors rallied against the Dakota Access Pipeline in North Dakota during 2016. "The similarity is [that] we are standing up to protect our water. And we are going to do this in a peaceful way," George said. "It is going to mark a day in history; it will be a massive mobilization" (Mapes 2018).

The Canadian federal government approved the pipeline expansion in November of 2016. The pipeline was opposed by the province of British Columbia as well as the largest cities in the lower mainland: Vancouver, Burnaby, and Victoria. "All hell is going to break loose on March 10, and thereafter," said Murray Rankin, a Member of Parliament representing the city of Victoria. "This is only the beginning, this is just the start, this is Ground Zero, day one. And it is going to go on for a very long time" (Mapes 2018).

Washington State tribes, state agencies, and environmental groups also rallied against the pipeline because it would increase oil-tanker traffic by 700% across northern Puget Sound, a bath-tub like enclosure with only one small exit to the Pacific Ocean. With their usual disregard for geography and climate, the designers of this giant oil port proposed to site it at the eastern edge of this bath tub, amidst rocky islands and winter storms. For those who dismiss the possibility of a major oil spill north of Seattle, opponents of this proposed connection between oil pipelines and tankers had two words: "Exxon Valdez." Anyone who wants to know what place these words hold in the history of oil spills should google those two words plus "Alaska."

Jan Hasselman, staff attorney at EarthJustice's Northwest office in Seattle, the lead attorney for the Standing Rock Sioux against the Dakota Access Pipeline said: "We are undergoing a cultural shift with a growing awareness that the benefits of fossil fuel projects simply aren't worth the cost. Momentum is building as the same fight plays out in British Columbia, with more people, and government agencies joining the side of indigenous people" (Mapes 2018).

"Our spiritual leaders today are going to claim back Burnaby Mountain," said Reuben George, of the North Shore's Tsleil-Waututh Nation as the crowd marched to the steady beat of drums and chants in opposition to the project. "It's going to take gatherings such as this ... [to] make sure the environment is not laid to waste and taken away from future generations. This is what we stand for today," George said. According to an account by Nick Eagland in the *Vancouver Sun* (2018), "The peaceful march ended up just outside Kinder Morgan's Burnaby tank farm, where [Vancouver, B.C.] Metro First Nations began construction of a wood-frame "Watch House" along the pipeline route. The building is expected to be completed by...will serve as a base for project opponents on the mountain."

A much smaller pro-pipeline rally (about 250 people) gathered in Vancouver on March 10, 2018. Supporters wore "I ♥ Oil & Gas" shirts, and carried placards bearing messages such as "I'll fight to lay pipe" and "Bring jobs, not lawsuits." "Crude oil in Canada is what pays the rent for us as a country," said Stewart Muir of

Resource Works Society, a nonprofit funded by the Business Council of British Columbia. The Canadian government bought the Trans Mountain pipeline from Kinder Morgan in 2018 for $4.5 billion. By May of 2022, the pipeline continued to be built slowly, roughly east to west, with a hornet's nest of political and legal conflict awaiting it on arrival in Vancouver.

"It's the ongoing proverbial battle between oil versus water," said Grand Chief Stewart Phillip, an Okanagan aboriginal leader and president of the Union of B.C. Indian Chiefs, a political advocacy organization representing 118 bands and First Nations within British Columbia. Phillip was one of several hundred people who were arrested in 2014 when they tried to block test drilling by Kinder-Morgan.

8.3 Oil Ports, Pipelines, Salmon, Dams, Native Peoples: And Greenhouse Gases

The conflict over pipelines and oil ports is only one part of this struggle over the future of nature in the Pacific Northwest. Another very important element is salmon, which spawn on streams that begin in the mountains. The salmon, which migrate from the mountains into the ocean and back to the exact spots where they spawned, are an environmental marvel. To those who look at the land, water, and air as a giant vending machine, the fish, like the oil, are money in the bank. Thus, the genesis of a modern-day battle between two human points of view.

Don Gentry, chairman of the Klamath Tribes, and Emma Marris on March 18, author of *Rambunctious Garden: Saving Nature in a Post-Wild World* described another brewing confrontation in Oregon in the *New York Times* (2018):

> Each spring and fall in the old days, Chinook salmon swam up the Klamath River, crossing the Cascade Mountains, to Upper Klamath Lake, 4000 feet above sea level. For millennia, the Klamath, Modoc, and Yahooskin Band of Snake Indians fished salmon from the lake and the river. The Klamath had agreements with the downriver tribes—the Karuk, Hoopa, and Yurok among them—to let fish pass so that some could swim all the way back to their spawning grounds. After dams were built on the river starting in 1912, the salmon were blocked. Today the only *"c'iyaals hoches"* (salmon runs) are enacted by the Klamath Tribes, whose members carry [a] carved cedar salmon on a 300-mile symbolic journey from the ocean to the traditional spawning grounds to bring home the spirit of the fish.

Four gigantic hydroelectric dams along the Klamath River near Oregon's border with California were being prepared for the largest dam demolition in U.S. history as of this writing. The idea is to open the river for salmon runs, a key part of native

8.3 Oil Ports, Pipelines, Salmon, Dams, Native Peoples: And Greenhouse Gases

peoples' traditions and survival. For migrating fish, a large dam is usually a death sentence.

According to Gentry and Marrismarch, (2018): "There's a threat to the dream of a revitalized river—a project that would put a newly unobstructed Klamath at risk of contamination while simultaneously contributing to climate change, desecrating grave sites and trampling the traditional territory of the Klamath people. If you've read anything at all about the protests near the Standing Rock Sioux Reservation in North Dakota, you might be able to guess what that threat is. It's a pipeline" (2018).

The pipeline's owner, Pacific Connector Gas Pipeline LP, and the developer of the export terminal, Jordan Cove Energy Project LP, both were owned by the Pembina Pipeline Corporation, based in the Canadian tar sands country near Calgary, Alberta. They sound a familiar theme: the new pipeline will provide jobs and taxes to some of Oregon's most impoverished counties, while disturbing nothing. Well, almost nothing, according to the vending machine crowd.

Twice during the Obama presidency, the Federal Energy Regulatory Commission ruled that the pipeline was not in the public interest, saying that the companies had failed to demonstrate that the benefits outweighed the project's negative impacts. In March of 2017, Gary Cohn, director of the National Economic Council (who resigned in March of 2018), sought to re-open the case, saying "The first thing we're going to do is we're going to permit an LNG export facility in the Northwest." By that time, President Donald Trump had appointed four of the Federal Energy Regulatory Commission's five members, so approval seemed likely.

"As far as the Klamath people are concerned," Gentry and Marrismarch wrote, "This pipeline is a bad idea even if the price of gas [was] predicted to skyrocket. The Klamath people oppose this project because it puts at risk their watersheds, forests, bays, culture, spiritual places, homes, climate and future." They continued (2018): "The 95-foot-wide gash through the tribes' ancestral territory that pipeline construction would require would be likely to unearth long-buried ancestors and pulverize sites of cultural importance. Construction would strip shade from streams and pollute them with sediment, harming fish central to the Klamath's traditions and way of life. If the pipeline catches fire or leaks, the Klamath River and its fish will be put at risk. The track record of fossil fuel pipelines suggests such a calamity is only a matter of time."

An analysis by Oil Change International, which sponsors clean-energy research, said that the proposed pipeline would transport enough natural gas to emit 2.2 million metric tons of carbon dioxide, methane, and nitrous oxides and would become the state's single largest source of these greenhouse gases, becoming its biggest single source of those emissions. The Klamath, Yurok, and Karuk tribes joined hundreds of landowners, conservation groups, and other citizens, in legal actions that have stopped the Dakota Access pipeline (thus far), an alliance much like that in Nebraska against the Keystone XL. The sheer familiarity of environmental damage undergirds Native American peoples' fierce opposition to the Dakota Access Pipeline. The determined nature of this resistance across North America stems from Native Americans' long history as resource colonies which hold more than a third of United States E.P.A. Superfund acute pollution sites.

8.4 Reorientation Respecting Humans' Relationship Toward the Natural World

The fact that Native peoples are in the majority at so many protests of Earth's desecration is not an accident or a coincidence, but evidence of a basic perceptual differences in cultural view *vis a vis* humans' relationship toward nature.

You may have heard the story of a young Native person, probably male, who, in search of employment, found a chain saw and began cutting trees. Soon he quit because he heard the trees screaming. Whether this actually happened (in that it can be "proved" by the usual standards of scientific method) may be beside the point: that Native people raised within a certain view of nature (and human nature) are capable of hearing trees scream.

To understand such an orientation requires an adjustment, a reorientation respecting humans' relationship toward the natural world. This situation goes to the heart of protests, et al. by Native people against "development." It's not just a matter of having something to eat (although that is a factor). It is a cultural attribute. Native people with this naturalistic view of the universe believe that everything has an animus, even the rocks between one's feet, and certainly the trees that screamed at the woodcutter. People who hold these cultural attributes (which also have been learned by non-Indians) see works of respect by a mother earth, whereas those who do not usually see the world mainly as a source of exploitive resources—a mother lode. These differences are illustrated in a National Park Service statement, available at: http://npshistory.com/publications/glca/piapaxa-uipi.pdf

> Southern Paiute people express a preservation philosophy regarding traditional lands and the animals, plants, artifacts, burials, and minerals that exist there. This philosophy primarily derives from a supernaturally established relationship between these lands and the people who have lived there since creation. One holistic philosophy that logically derives from this human–land relationship addresses the issue of how to act toward the land, animals, plants, artifacts, and burials. Simply, the philosophy leads to the normative assertion that these cultural resources should be left undisturbed, i.e. they should be preserved as they are, not removed or modified in any way.
>
> This philosophy is in sharp contrast with an instrumental human-land philosophy that leads to the normative assertion that the land, animals, plants, artifacts, minerals, and even burials should be used for economic development or scientific study. This philosophy is premised on the epistemological belief that humans should dominate and control the natural environment for the immediate benefit of whomever is sufficiently powerful to hold sovereignty over the land.

(continued)

8.4 Reorientation Respecting Humans' Relationship Toward the Natural World

> Consistent with this instrumental human-land philosophy, unused natural or human resources have a potential for development, being termed "wild lands" or "wild people," and therefore constitute a challenge for development efforts. The process of conquering wild resources has variously been termed "progress," "modernization," "civilization," and "development." The human and natural components of a system are perceived of as "resources" to be "managed."
>
> The very terms used to describe the scientific study of these resources—"-American Indian Cultural Resources"—reflect a philosophy that is antithetical to the core philosophical belief of Southern Paiute people. Understanding the existence of a conflict in basic philosophies is a complex, but essential, starting point for explaining why the Southern Paiute people have made certain types of cultural resource responses.

Concerns for plants that Southern Paiutes rely upon for survival made ethnobotanical knowledge essential to their "transhumant adaptive strategy" for living in the desert. An intimate knowledge of plant genetics has been suggested as a major "cultural focus" of desert-dwelling Indian people. Being horticulturalists is a cultural characteristic that separates Southern Paiutes from closely related groups in the Great Basin. A wide variety of plants continue to be utilized by Paiute people for food, medicine, ceremonies, and economic activity.

It is evident that plants are important in Paiute cosmology because Paiutes recite a prayer before a plant is picked and utilized with a request that it provide needed medicine or nutrition. The plant, like the people, has rights and human-like qualities. The prayer is directed to the plant because it is perceived as an anthropomorphic organism. Indigenous people express concern for all animals and plants because of a traditional belief that everything, whether considered animate or inanimate in non-Native belief systems, including insects, rocks, et al. are important to the earth. Respect for animals is demonstrated by the kinds of traditional prayers that are recited in association with hunting that includes the taking the life of an animal. Plants, like animals, are perceived to have rights and human qualities because they are seen as relatives to human beings. Birds like eagles are perceived as important and are prayed to and talked with when captured.

The Indian people[s] believe natural elements should be protected from contamination, alteration, and even movement without talking to them. Like plants and animals, Southern Paiutes believe that natural elements have rights, human attributes, and lives of their own. The belief that the water sources are connected to each other underground correlates with the belief about Water Babies. Southern Paiute people mentioned that Water Babies are often present at springs. According to the ethnographic literature, Water Babies owned springs and had elaborate systems of underground pathways, usually taking form as underground watercourses. The Water Baby used to travel from one spring to another. Water Babies are never good, at best being neutral, and are extremely dangerous. If a person angers a Water

Baby, the person will almost surely die. By extension, then, any activity that damages or destroys the underground water sources will anger the Water Babies who own it, thereby endangering everyone in the vicinity.

8.5 Quapaw: Too Toxic to Clean up

Zinc and lead were mined within the jurisdiction of the Quapaw in and near Picher, Oklahoma until 1967, when, according to a retrospective by Terri Hansen for the Indian Country Today Media Network, "mining companies abandoned 14,000 mine shafts, 70 million tons of lead-laced tailings, 36 million tons of mill sand and sludge, as well as contaminated water, leaving residents with high lead levels in their bodies. Cancers skyrocketed, and 34 percent of elementary-school students suffered learning disabilities" (Hansen 2013).

The Quapaw homeland once encompassed much of Arkansas. They were removed to a small tract of land during the 1830s, and whites later moved in around them. Once a nation of at least 35,000 people, the Quapaws were reduced to about 150, mainly by smallpox and other diseases. They were battling back when the mining ruined their new homeland.

By the late 1960s, roughly 178 million tons of chat (waste), mill sand, and sludge in about 30 piles dominated the landscape in and near Picher. Children played on the chat piles and members of the high school's track team trained on them, soaking up the same lead, zinc, cadmium, arsenic, iron, and manganese that were turning local drinking water a shade of rusty orange. Families laid out picnics on the chat piles and fathers built sandboxes filled with the gray sandy material that, unknown to them, was poisoning their children. Children became sick, and lagged behind in school, as doctors reported unusually high rates of several cancers, hypertension, respiratory infections, and high infant mortality rates. The human devastation was confirmed in 2003 when the rate of the chronic lung disease pneumoconiosis was reported to be 2000% higher than normal.

Karen Harvey, 53, who lived in Picher from 1960 to 2002, said that children played on the chat piles and swam in tailings ponds on hot days. "We'd go swimming in them and our hair would turn orange and it wouldn't wash out," she said (Shephard 2014).

The area was designated as the Tar Creek Superfund Site in 1983, but after Picher was initially scheduled for cleanup, but later assigned it to a special environmental hell: too toxic to allow continued habitation. The federal government initially spent $138 million replacing soil around many houses, an average of $70,000 per house in an area where the homes themselves averaged $58,000 in value. All told, the remediation effort did very little to address systemic problems caused by leaks from the chat piles and massive water pollution. One report estimated that 50 large dump trucks would be required around the clock for 40 years to clear the chat piles—and then where would they dump it?

After that first failed cleanup effort, the federal government offered a rock-bottom buyout that paid people who now owned houses that had been reduced to worthlessness to leave town. The federal government bought out 2000 properties in Picher, and pledged that the 103-square-mile site's soil and water would eventually be detoxified and converted into a wetland, a process that could require several decades.

Anar Virji wrote for Aljazeera in May 2014 that "From a distance, they look like mountains sloping across the Oklahoma plains. But as my reporter, cameraman, and I get close, we see [that] they're massive piles of tiny rocks: millions of tons of toxic waste called "chat" that came from the abandoned lead mines that make up the site of Tar Creek. This place was once booming, home to small mining communities. Shells of houses still stand in what was once the town of Picher, a reminder that it used to be home to close to 15,000 people at its peak."

Even in 1967, when the mines closed, leaving behind huge mounds of waste, the people of Picher and surrounding areas had no idea of their toxicity, as pollution leached into the ground, turning water and soil a rusty orange hue, seeping into water wells and the pipes of houses. "The groundwater ran into the wells, it had lead content in it," said Ranny McWatters, treasurer of the Quapaw tribe. "Over the years, it got into the children" (Virji 2014). By 1993, tests revealed that a third of children had elevated levels of lead in their blood, negatively affecting development of their brains.

Until it is cleaned up, if ever, much of the Quapaw homeland will be nearly uninhabitable. On March 25, 2013, the Quapaw sued the United States federal government for $75 million, alleging failure to clean up what remains of the United States' largest lead and zinc mine, which had produced billions of tons of ore. Much of the reservation's water had become severely polluted. Bullfrog Films and director Matt Myers made a documentary film on the Quapaw pursuit of justice, released in 2011. The film, simply titled *Tar Creek,* provides a searing portrait of human and natural damage wrought upon a 47-square-mile area in northeastern Oklahoma now known as the Tar Creek Superfund site, where residents are still seeking relief from acidic mine water in creeks, lead poisoning of children, and sinkholes that swallow backyards of contaminated houses. After the town's site had been emptied and cleanup began, the site was devastated by a major tornado.

The film follows people's responses as the government conducted a slow-motion buyout in which very few people received anything close to what they would need to start over elsewhere. The camera pans a dead landscape largely bereft of animal, fish, and bird life as human communities slowly disintegrate. One resident compares his community's fate to Rachel Carson's *Silent Spring*. He is wrong, in a sense, because this environmental apocalypse—forty-seven square miles of prairie turned into permanent wasteland—is worse than anything described in Carson's book.

In 2014, Dan Shepard of NBC News wrote that "Thirty years and hundreds of millions of dollars since work began to clean up this former lead and zinc mining boomtown, progress is still being measured in inches or feet. That gauge is provided by the towering but slowly diminishing piles of 'chat,' or tailings, that still loom over what is now a virtual ghost town" (Shephard 2014). Shepard wrote that "Today, there are only about 10 hardy souls left in Picher to watch as the cleanup slowly carves away at the dozen or so remaining piles of chat and restores Tar Creek, whose waters still run red from heavy-metal runoff, to some semblance of its former self. If their health holds out, they likely have another 20 or 30 more years to take in the slow-motion show in this quiet northeastern corner of Oklahoma, near the Kansas and Missouri borders." In 20 years, the federal government had spent more than $300 million there, and the job was only partially done in an area where pollution was worse than that of the infamous Love Canal in Upstate New York.

Workers removed some of the toxic sludge as they refilled streambeds with clean soil and reduced chat piles. Picher was only a small part of the area requiring eventual cleanup: the "Tri-State Mining District" that stretches over 2500 square miles from Oklahoma into parts of southwestern Missouri and southeastern Kansas, where soils and waters have been polluted by mining wastes.

Jack Greene, 91 years of age in 2014, worked in the mines under Picher between 1938 and 1941 (when he was drafted into the Army for World War II) told Shepard that tremors from the mines shook pencils off desks at the local high school. "They let people destroy this country from being so greedy," he said (Shephard 2014).

By mid-2014, parts of the 40-square-mile Tar Creek Quapaw toxic mining site, most notably areas called the Catholic Forty and Chat Base 11 North were being ever so slowly cleaned up as part of the first Superfund project by the Quapaw tribal government, cooperating with state and local environmental agencies. "We completed the first cleanup less expensively and better than previous efforts," said Quapaw Chairman John Berrey. "Our goal is to make this land useful and productive again. We live here and we care about the outcomes, so we are very pleased to have these two new agreements in place" (Quapaw Take Lead 2014). Roughly 72,000 tons of contaminated material had been removed from the Catholic Forty by 2014— a start, with a lifetime of decontamination to come.

Even as some of the area was being cleaned up, Picher's slow-motion demise continued for many years, as the underground mines, no longer propped open from below, caused the surface to collapse into sinkholes to a degree that an Army Corps of Engineers study released in 2006 indicated that almost 90% of the town's buildings had become subject to sudden collapse. Occasional earthquakes that have been attributed to fracking (removal of underground oil and gas along with large volumes of contaminated water) accelerated the collapse of Picher.

8.6 Coal: Pacific Northwest Tribes Protest Transport

For fossil-fuel firms in North America with an eye on burgeoning markets in China and India, the shortest line between profit and the coal fields of Wyoming, the oil deposits of the Bakken formation in the Dakotas, and the tar sands oil of Alberta is the Pacific Northwest, most notably the Salish Sea, a strip of bays, straits, islands, and inlets between Vancouver, B.C. and Seattle. Native peoples in that area, allied with several environmental groups by 2022 seemed to have fought plans to ship coal and oil to the Pacific coast and Columbia to a standstill. What remained was Cherry

Point, in very northernmost Puget Sound, a few miles south of the Canadian border, already the site of a large oil-refining complex.

Plans for a terminal at the Port of Grays Harbor, near the Quinault reservation on the Pacific Ocean coast of Washington State were shelved in 2012. In Oregon, three proposals for coal-export terminals met stone cold receptions. Morrow Pacific in St. Helens also ran into Native American and environmental resistance; plans for a coal terminal at Port Westward were abandoned by Kinder Morgan in 2013, and Project Mainstay, proposed for the Port of Coos Bay lost most of its investors at about the same time. Yet another one, The Millennium Bulk Terminal in Longview, Washington was still being pursued.

Cherry Point is also a favored fishing ground of the Lummis, whose reservation lies north of Bellingham, Washington, and south of Vancouver, British Columbia. The Lummi adamantly opposed plans for the Gateway Pacific Terminal, a plan to export 48 million tons of coal from eastern Montana and Wyoming's Powder River Basin, averaging nine coal trains a day. They saw the terminal as a long-term threat to their fishing, a 2500-year old tradition, and today a multi-million dollar annual harvest of salmon, crab, and halibut, basic to their *schelangen*, ("*how life ought to be lived*").

Pacific Terminals, a subsidiary of SSA Marine, is among the largest shippers of coal on Earth. Amber Cortes, writing in *Grist* during 2014, outlined the battle shaping up:

> At Cherry Point, the site of the proposed coal terminal, the water is great for fishing: the depth of the sea bottom drops off quickly from the shore and attracts fish like surf smelt, sand lance, and herring that come to feed and spawn. It's one of the reasons Cherry Point is so special, and a state aquatic reserve. That very deep water, so close to shore, is what makes Cherry Point so special to Pacific Terminals and SSA Marine, too. You can't bring tankers and bulk carriers that draft in excess of 50 feet deep into just any port. "But what's good for the shipper is critical for the fish," said Fred Felleman, a marine biologist who consults with Pacific Northwest tribes and environmental groups about shipping impacts on the marine environment.

Many people in the area who had expressed concern that coal shipments threaten the area's natural bounty recalled a coal-ship spill in 2012 in Vancouver at Westshore Terminals. The ships that carry coal are bigger than oil tankers, and tougher to steer though rock-studded inlets. They also are not legally required to employ tugboats as escorts. The upshot, according to the Indian-environmentalist alliance, is a risky future on the Salish Sea, a complex marine environment full of islands that is dense with other shipping traffic.

Pacific Terminals made no friends in June 2011, when it bulldozed wetlands at Cherry Point without environmental permits. The same area was registered as a state archeological site containing 3000-year-old human remains (Cortes 2014). Soon the

Lummi were burning symbolic checks in front of television cameras with this message: No amount of coal money would buy them off. Totem poles were fashioned by master carver Jewell James in protest of the coal terminal, to be carried eastward across the United States, starting at the desecrated earth of Cherry Point:

> The totem pole will stand at 19 feet tall when completed. Carved into the pole is Mother Earth, lifting a child up to receive the teachings of the heavens. On the bottom of her dress are four warriors who are guardians of the sacred earth-power, and a snake, representing the power of the earth traveling upwards (Cortes 2014).

James said that the totem pole "depicts a woman representing Mother Earth, lifting a child up; four warriors, representing protectors of the environment; and a snake, representing the power of the Earth" (Walker 2014). It was carried to the source of the coal, as well as the tar sands. The Lummis, in the meantime, assert that the coal terminal if built will impede their rights under the Treaty of Point Elliott (1855), as upheld in U.S. federal courts during the 1970s. This argument already has been used to persuade the Army Corps of Engineers deny permission to build a commercial salmon fish farm in the same area. Protection of Quinault fishing rights was a major factor in the defeat of two proposed oil terminals at Gray's Harbor.

A protest of coal transport to the Gateway Pacific Coal terminal assembled at Cherry Point, Washington (which the Lummi call *Xwe'chi'eXen*), led by Native fishing people, with non-Indian allies. "We have to say 'no' to the coal terminal project," said Cliff Cultee, chairman of the Lummi Nation. "It is our *Xw' xalh Xechnging* [sacred duty] to preserve and protect all of *Xwe'chi'eXen* [the Lummi homeland]" (Northwest Tribes 2012). The Indian Country Today Media Network reported that "During the protest, the tribes held a ceremony of thankfulness, remembrance and unity to solidify their commitment." The statement noted that Lummi Indians have the U.S.'s largest Native fishing fleet and have fished near Cherry Point for thousands of years.

Writing August 11, 2014 in the *Bellingham Herald*, Jewell James asserted that the coal terminal proposed at Cherry Point poses "a tremendous ecological, cultural and socio-economic threat" to Pacific Northwest indigenous peoples. "We wonder how Salish Sea fisheries, already impacted by decades of pollution and global warming, will respond to the toxic runoff from the water used for coal piles stored on site." James continued: "What will happen to the region's air quality as coal trains bring dust and increase diesel pollution? And of course, any coal burned overseas will come home to our state as mercury pollution in our fish, adding to the perils of climate change." Marianne Maumus of Ochsner Health Systems told the New Orleans *Times-Picayune* that coal dust contains heavy metals including arsenic, cadmium and mercury, and can cause cancers, neurological, renal and brain-development problems. "I think the risk is real. I think there is a lot of potential harm from multiple sources," Maumus said (Walker 2014).

8.6 Coal: Pacific Northwest Tribes Protest Transport

The Affiliated Tribes of Northwest Indians (ATNI) passed a resolution in September 2012 requesting a full environmental analysis by the U.S. Army Corps of Engineers of several coal-transport proposals for terminals on the Columbia River's Port of Morrow and Longview, as well as at Cherry Point north of Bellingham near the Lummi Nation (near an existing oil refinery). "This is not about jobs versus the environment," said James, "It is about what type of jobs are best for the people and the environment" (Northwest Tribes 2012). Despite the Lummis' strident opposition to this project, its sponsors continued to knock on their doors through at least 2022 (when this book's text was finished), as Lummi Nation Chairman Tim Ballew pointed to a number of accidents involving coal and oil-carrying rail cars across North America during the previous several years, and said:

> "I can assure you that we have carefully considered the impacts associated with this project and have concluded that these impacts simply cannot be avoided, minimized or mitigated," Ballew wrote. "While we appreciate your desire to engage on these issues, we remain steadfastly opposed to this project and do not see the utility in pursuing any further discussion. Please be advised that we fully intend to ensure that our treaty rights are honored and that the spiritual and cultural significance of *Xwe'chi'cXen* is protected" (Walker 2015).

In August 2014, the Oregon Department of State Lands rejected a coal-terminal application along the Columbia River, citing Native treaty and fishing-rights considerations. Local tribes celebrated, and others to the north, in Washington State, saw the decision as a possible precedent-setter for fossil-fuel terminals that they were resisting. The department said that "while the proposed project has independent utility, it is not consistent with the protection, conservation, and best use of the state's water resources, and that the applicant did not provide sufficient analysis of alternatives that would avoid construction of a new dock and impacts on tribal fisheries. Ambre Energy, which had applied to ship 8.8 million tons of coal a year to Asia via the Port of Morrow's Coyote Island Terminal." The company said it would appeal.

"Today's landmark decision reflects what is in the best interest of the region, not a company's pocketbook," said Carlos Smith, chairman of the Columbia River Inter-Tribal Fish Commission. "This decision is one that we can all celebrate. It reaffirms the tribal treaty right to fish, is in the best interest of the Columbia Basin's salmon populations, and our communities. It is a reflection of what is best for those who would be forced to live with the consequences of Ambre's proposal, not what is best for those who would profit from it. This is the beginning of the end for this toxic threat–the Tribes will stand with the State to protect its sound decision" (Treaty Victory 2014).

"The State's action makes a strong policy statement by recognizing Tribal Sovereignty and the Treaty Rights of the Columbia River tribes," said Lummi Nation Chairman Timothy Ballew II. "Such decisions are few and far between.

This is important not just for the Yakama and Umatilla but all Indian fishing tribes. Together we can, and will, protect our way of life." Yakama tribal council chairman JoDe Goudy, through whose territory trains would pass en route to a terminal on the Columbia River, said: "[The] Yakama Nation will not rest until the entire regional threat posed by the coal industry to our ancestral lands and waters is eradicated" (Treaty Victory 2014).

Native leaders in the Pacific Northwest have continued to rally against plans to transport coal from the Pacific Northwest to Asia through waterways whose fish and shellfish sustain their lives, asserting, according to the Indian Country Today Media Network, that "Treaty fishing rights are meaningless if there are no healthy fish populations left to harvest" (Hansen 2012, August 15). Since the Boldt Decision (1974), which allocated up to half the salmon in much of Western Washington to Native fishing people, Native peoples have worked with state authorities to sustain the runs. Coal transport threatens that effort. Coal dust from trains may degrade human health, as well as that of wildlife and fish. The companies are planning to add rail capacity to the single line that was transporting coal over the Cascades and Rockies from Wyoming's and Montana's Powder River mining country. Trains carrying coal from Wyoming reach a mile in length, with plans to double that capacity.

The Lummi Nation wrote to the Army Corps of Engineers July 30, 2012: "Any impact on Lummi treaty fishing right is inherently an impact on the Lummi way of life," wrote Lummi Indian Business Council Chair Tim Ballew in the letter. "We believe that the Corps should see that these projects would without question result in significant and unavoidable impacts and damage to our treaty rights" (Lummi Nation 2012). The letter said that "The Lummi Indians maintain the largest Native fishing fleet in the United States. Moreover, Lummi fishers have worked in the *XweChiexen* [lLummi homeland's] fishery for thousands of years. The Gateway Pacific export terminal would be the largest such terminal on Turtle Island's [North America's] west coast....It would significantly degrade an already fragile and vulnerable crab, herring and salmon fishery, dealing a devastating blow to the economy of the fisher community," the tribe's statement said (Lummi Nation 2012).

The Public Broadcasting System (P.B.S.) "News Hour" took up the Lummis' resistance to the Cherry Point coal terminal proposal on August 2, 2012, during which health professionals described the dangers of coal dust and diesel exhaust, as Lummi fishing people spoke of threats to their harvest. As proposed by SSA Marine in 2012, the coal terminal would export 54 million metric tons per year of "bulk commodities," including 48 million metric tons of coal. "If the Lummis come to that position, it will make us reassess the direction we are going," said Muffy Walker, the head of the Army Corps of Engineers' regulatory branch in Seattle. "We have denied permits in the past, based on tribal concerns" (PBS Newshour 2012).

Opponents of the proposed coal-export terminals decried the noise, dirt, and smell of the trains, and said that coal dust may threaten herring, Chinook salmon, and killer whales. Environmental problems already have reduced salmon runs to the lowest since Judge George H. Boldt's landmark decision in 1974, according to a report from the Northwest Indian Fisheries Commission. The ICTMN reported: "Many worry that such a network of coal transport would degrade salmon and cultural-foods habitats as well as affect treaty rights. Among the concerned groups are the Columbia River Inter-Tribal Fish Commission (CRITFC), the National Wildlife Federation (NWF) Tribal Lands Program, and tribal nations like the Lummi in northwest Washington and the Yakama in eastern Washington" (Hansen 2012).

"Our treaty rights to first foods—salmon, deer, chokecherries, mountain sheep, wild roots and other resources—are retained by the tribes, and they weren't given to us by the government," said Bruce Jim, a Confederated Tribes of Warm Springs elder on the tribe's Fish and Wildlife Committee. "Moving coal through our land will threaten these rights and our connections to these plants and animals" (Hansen 2012). On the coast, a report, "The True Cost of Coal," by the NWF, anticipates that "habitat degradation from port expansions and shipping traffic; decreased water quality from coal dust; mercury deposits from coal burning and wind-driven transport, increased carbon pollution and ocean acidification" (Hansen 2012).

Peabody and other coal companies had been planning to ship coal, as much as 150 million tons a year, by river, rail, and road, mainly to China. By 2013, coal consumption and prices in the United States were in decline. Electric utilities in the United States used 18% less coal in the first half of 2012 than during the first half of 2011. Coal consumption in 2012 was 27% below that of its peak year, 2008, following several years of steady growth. Coal is being replaced for electrical power generation by natural gas from "fracking," which produces about half the carbon dioxide per energy unit as coal—it's still a fossil fuel, but a cleaner one. Coal is now being mined for export to Asia (mainly to China), requiring West Coast terminals in Western Washington and along the Columbia River, where Indian tribes have treaty rights for salmon.

"The idea of...new coal export terminals in Western Washington and Oregon—and the hundreds of trains and barges running from Montana and Wyoming every day to deliver that coal—would threaten our environment and quality of life like nothing we have seen before," said Billy Frank Jr., chairman of the Northwest Indian Fish Commission. "Coal may be a cheap source of energy for other countries, but these export facilities and increased train traffic would come at a great cost to our health, natural resources and communities" (Trahant 2012). Swinomish Tribal Chairman Brian Cladoosby said: "If a coal train or tanker were to spill on the route or in the river at Port Morrow in Oregon, the waterways will carry the pollution throughout the Northwest, and coal dust will be carried through the mountains in the air we all breathe" (Trahant 2012).

8.7 Coal Transport and the Global Greenhouse Gas Load

Native peoples pointed out that the coal-export business is being subsidized by the U.S. federal government. The Institute for Energy Economics and Financial Analysis reported in 2012 that coal mined on lands owned by that government is so cheap (less than $10 a ton) that taxpayers are losing $1 billion a year in royalties, as the companies are able to sell what they mine for as much as ten times what they have paid. "Taxpayers are likely losing out so that coal companies can reap a windfall and export that coal overseas, where it is burned, worsening climate change," said U.S. Senator Edward J. Markey, a Massachusetts Democrat (Riordan 2013, A-23).

Carbon dioxide emitted by burning coal adds to the global burden of greenhouse gases that add to global warming. According to one analysis, "Projected exports from Cherry Point alone could result in more than 100 million metric tons of carbon dioxide annually. Added CO2 has already begun to boost the acidity of near-shore waters, threatening Washington's shellfish industry. The billions of tons of coal burned in Asia every year contribute markedly to global warming. Should the United States be selling them subsidized coal and encouraging this impending disaster?" (Riordan 2013, A-23).

Residents of the Northern Cheyenne Reservation in eastern Montana also objected to coal transport, asserting that "coal mining and the construction of a railroad to transport the coal would devastate their essential Otter Creek and the Powder River Basin" (McNeel 2012). The Northern Cheyenne allied with ranchers, hunters, and conservation groups such as the National Wildlife Federation against the new coal railroad "and its ruinous effects on its multitude of wildlife, cougars, black bears, elk, mule deer, white-tailed deer and pronghorns all roam the forests and sagebrush prairies that make up this portion of the Powder River Basin in the easternmost portions of Montana and Wyoming. Otter Creek, that supports more than 20 fish species, empties into the Tongue River and eventually the Yellowstone River" (McNeel 2012).

Local residents vented opposition in five public hearings during November 2012. "Almost 100 percent of the people spoke out against it," said Alexis Bonogofsky, a Northern Cheyenne tribal lands program manager for the National Wildlife Federation in that region. "It was conservative ranchers and liberal ranchers. There were cowboys and Indians. Everyone was just saying, 'We don't want this train.' It was remarkable because in some of those communities, like Forsyth and Ashland, the perception before this was that they supported it." (McNeel 2012).

On December 4, 2012, another hearing was attended by about 800 people in Spokane, Washington. In Seattle, about 2000 people attended a similar hearing December 13, 2012. The crowd in Spokane included supporters of the trains and terminals, such as railroad workers, but about 75% of the people who spoke were opposed, citing "dangers posed by railroad crossings, slower emergency-vehicle response time, potential coal-dust-related health problems and other concerns. Some came all the way from southeastern Montana, including nine people from the Northern Cheyenne Reservation who made the nearly 1,300-mile round-trip" (McNeel 2012).

"I'm transitioning into something that is a better vision for the future, more involved in renewable energies that should be embraced that we, as global citizens, should embrace," said Jeff King, Northern Cheyenne, who had been working in the coal industry. "I would like to see our nation embracing renewable energy" Burdette Birdinground, a Crow with Northern Cheyenne relatives, drove from the Flathead Reservation. "It will pollute the waters, the air, will ruin the cultural sites in the valley and the hunting grounds," he said. Vanessa Braided Hair said: "stop the destruction of our Mother Earth" (McNeel 2012).

8.8 The Lubicon Cree: Land Rights and Resource Exploitation

Despite the fact that more than $20 million worth of oil has been pumped out of their land since 1980, the 500-member indigenous Lubicon Cree community of Little Buffalo, in far northern Alberta, has no running water, inadequate housing, no sewage, and no public infrastructure (Green 2005). The Lubicons have been faced with a choice: rely on an overtly hostile provincial government or move in with another First Nation that can offer them social services. In the words of Chief Bernard Ominayak, the proposed change could "tear our people apart" (Unions 2000). In 1990, after six years of deliberation, the United Nations charged Canada with a human-rights violations under the International Covenant on Civil and Political Rights, stating that, "recent developments threaten the way of life and culture of the Lubicon Lake Cree and constitute a violation of Article 27 so long as they continue" (Resisting Destruction 2017).

The Lubicon land claim remains outstanding, as the government of Alberta continued to lease Lubicon territory to multinational corporations that had exploited and contaminated the land. The toll on peoples' health is startling. In 1985 and 1986, of 21 Lubicon pregnancies, 19 resulted in stillbirths or miscarriages.

In 1971, the province of Alberta announced plans for construction of an all-weather road into the Lubicons' traditional territory to provide access for oil exploitation and logging. Road-building plans were undertaken without Lubicon consent and resisted by the native band in Canadian courts. At one point, the federal government asserted that the Lubicons were "merely squatters on Provincial crown land with no land rights to negotiate" (Resisting Destruction 2017).

Having cleared legal challenges, Alberta built the all-weather road into Lubicon territory during 1979. Construction of the road was followed by an explosion of resource-exploitation activity that drove away moose and other game animals, causing the Lubicon traditional hunting-and-trapping economy to collapse. Within four years, by 1983, more than 400 oil wells had been drilled within a 15-mile radius of the Lubicon community. From 1979 to 1983, the number of moose killed for food dropped more than 90% from 219 to 19, as trapping income also dropped 90% from $5000 to $400 per family. At the same time, the proportion of Lubicon Cree on welfare shot up from less than 10% to more than 90%. During 1985 and 1986, 19 of 21 Lubicon pregnancies resulted in stillbirths or miscarriages. "In essence, the Canadian government has offered to build houses for the Lubicon people and to support us forever on welfare—like animals in the zoo who are cared for and fed at an appointed time," said Chief Bernard Ominayak (Resisting Destruction 2017).

The Lubicon Lake Cree have a claim to about 10,000 square kilometers of land in northern Alberta, east of the Peace River and north of Lesser Slave Lake. The land was regarded as so remote in 1900 that Canadian officials seeking to negotiate treaties completely ignored it. The Lubicon did not sign Treaty 8, which was negotiated in 1899 and 1900, providing the band no legal title to its land—title they were still seeking to negotiate with Canadian officials more than a century later.

During the century that the Lubicons have sought legal title to their land base, it has been scarred by oil and gas production, as well as industrial-scale logging. While Canadian officials have refused to set aside land for the Lubicons, large oil and gas, pulp and paper, and logging companies have moved into the area to exploit its natural wealth (Green 2005).

The Lubicons were first promised a reserve by the Canadian federal government in 1939, before oil was discovered under their lands. Oil companies flooded into the area in the 1970s, all but destroying the Lubicon society and economy. The story of the Lubicons is a case study of how modern resource exploitation can ruin a natural setting and the indigenous people who once lived there.

The Lubicons engaged in negotiations with the federal and provincial governments three times during the 1990s, but the talks broke off because the two sides failed to agree on the size of the land mass and monetary compensation that should be allotted to the Lubicon (Green 2005). "It is very worrisome when you are at the table year in, year out…with government sponsored and supported resource development…subverting the rights you are at the table to negotiate." "You have to wonder if there is any sincerity about achieving a settlement," said band advisor Fred Lennarson (Guerette 2001).

During the middle 1980s, after the provincial government enacted retroactive legislation to prevent the Lubicon from filing legal actions to protect its traditional territory from booming oil and gas development, the Lubicon launched a protest campaign against petroleum companies sponsoring 1988 Calgary Olympics (Guerette 2001). On the eve of the international Olympic event Premier Don Getty established a personal dialogue with Lubicon Chief Bernard Ominayak which lead to the Grimshaw Accord, an agreement committing the province to transfer to Canada the 95-square-mile reserve the Lubicon had been seeking (Guerette 2001).

Oil and gas revenues from Lubicon ancestral lands continued at about $500 million a year. Not a penny went to the Lubicons. During 1988, after 14 years getting nowhere in the courts, the Lubicons asserted active sovereignty over their land. A peaceful blockade of access roads into their traditional territory stopped all oil activity for six days. The barricades later were forcibly removed by the Royal Canadian Mounted Police. Alberta's Premier Don Getty then met with Lubicon leaders in Grimshaw, Alberta; the result is an agreement on a 243 square kilometer (95.4 square mile) reserve area called the "Grimshaw Accord."

Twelve years later, the provincial government abandoned the agreement. In the meantime, environmental assessments disclosed that more than 1000 oil and gas well sites had been established within a 20-kilometer radius of Lubicon Lake, on land that had been promised to the indigenous people.

8.9 Logging on Lubicon Land

Forest-industry companies by 2001 held concessions from the provincial government that covered nearly all of the land claimed by the Lubicon not already leased for oil and gas production. The first of the industrial loggers, the Japanese paper company Daishowa began logging Lubicon land during the 1980s. In 1988, Daishowa announced plans for a pulp mill near the proposed Lubicon territory that would have processed lumber equaling the area of 70 football fields daily. The province of Alberta also granted Daishowa timber rights to an area including the entire Lubicon traditional territory.

An international boycott of Daishowa was launched to protest the company's clear-cutting of Lubicon land. In response, Daishowa stayed off Lubicon land during the 1991–1992 winter logging season. Logging later resumed, as the company sought to lift the boycott by suing Lubicon non-Native allies in Canadian courts. Daishowa's legal action was thrown out of court during the late 1990s, as the boycott intensified. Daishowa, which manufactures paper bags, newsprint, and other paper products, retracted its plans to log Lubicon land only after the boycott began to reduce its revenues. Daishowa then again pledged to stay out of Lubicon forests until land rights were delineated. At this point, the boycott ended. The company asserted that the boycott had cost it $20 million in lost sales (Guerette 2001).

8.10 Conflicts over Resources Continue

During May 2011, the Lubicon village of Little Buffalo was hit by the largest oil spill in Alberta since 1975 when more than 28,000 barrels leaked from an aging pipeline and contaminated rivers and the water table. The Forum on Religion and Ecology at Yale University reported that "Members of the Lubicon community are reporting illnesses as a result of the stench of oil in the air, and that the company involved is not providing clear information....Some 20 local people have been hired to help with the 24-hour a day cleanup operation, but the spill has brought into sharp focus some of the long-standing ills that afflict Little Buffalo. The community has no running water; instead, water is trucked in, and people hand-carry it into their houses from 45-gallon oil drums. With no plumbing, people rely on outhouses. Showers are available at the school, but many wash using sponge baths" (Oil Spill 2011).

By 2014, Lubicon Cree dissidents were maintaining a campaign against fracking on their land, with signs such as "Frack Off! This is Lubicon Land!" As most people at Little Buffalo went without running water, one chief received Canadian $1.5 million through the Cree Development Corporation (CDC), which is technically non-profit, but generates millions of dollars through contracts with the energy industry. It was not clear, due to sloppy accounting, whether these were payments for personal use, or if they were "issued because the directors were buying construction equipment so they could bid on jobs" (Lubicon Chief 2014).

"Our infrastructure is basically non-existent," said Chief Billy Joe Laboucan. A Canadian Broadcasting Corp. (CBC) report said that "Laboucan was voted in as chief of the Lubicon Lake band in 2013, in the first federally recognized election for the band. He commissioned an audit to find out where all the revenue generated by CDC went. The auditor went through thousands of checks and other bank records and found that over a four-year period the directors of the CDC paid themselves close to $3 million. One of those directors was former chief and longtime leader Bernard Ominayak, who received 99 payments totaling $1.5 million." The report said that "living conditions are deplorable" in Little Buffalo (Lubicon Chief 2014).

Cheryl Ominayak, who has lived in Little Buffalo, a village of about 500 people, for 28 years, said she draws water from a barrel and uses an outhouse. "Summer bugs are bad but you have to use it—no running water," Ominayak said of the outhouse. "We are probably about the only community in northern Alberta that doesn't have running water" (Lubicon Chief 2014).

The assertion that the money was used for business expenses was received with skepticism by many community members, the CBC reported. "Shame on them, shame on them. I think everyone around this community deserves an apology," said community member Denise Ominayak. Cheryl Ominayak also wanted answers. "Where did they put the money? They obviously didn't give it to us community members because we all live in old mouldy, rotten houses" (Lubicon Chief 2014). In a rare move, Alberta's government provided trailers with water tanks and septic systems.

8.11 The Moapa Paiute: Goodbye Toxic Ash: Solar in, Coal Power out

Quite a few people who would like to retire smoke-belching, sulfur-spewing, carbon-pumping, coal-fired power plants might consider taking lessons from the Moapa Paiutes, a tiny American Indian tribe who live a few miles, as the toxic ash flies, north of Las Vegas, Nevada. The Moapa Paiutes, after decades of suffering the suffocating ash clouds of an old coal-fired plant not only retired it, but convinced the Interior Department to site two solar-power arrays nearby, creating jobs.

In 2013, the Reid Gardner Generating Station, a coal-fired power plant adjacent to the Moapa River Reservation (about 50 miles north of Las Vegas, Nevada), was slated for closure in 2017. The closure was good news, but they had a new problem: the Moapa Band of Paiutes had a new problem: ramping up pressure on its owner, NV Energy, Inc., to clean up residual pollution. On August 8, 2013, the band and the Sierra Club filed suit in U.S. District Court, Las Vegas, Nevada, to legally compel the cleanup. MV Energy was bought out in May, 2013 by Warren Buffet's MidAmerican Energy Holdings, which had taken an active role in raising Iowa's use of wind power for electrical generation in Iowa to one-half of its production by 2015. By 2022, Iowa was nearing 100% energy generation by wind.

8.11 The Moapa Paiute: Goodbye Toxic Ash: Solar in, Coal Power out

Reid Gardner was the last coal-burning power plant in Nevada. The lawsuit claimed "that the federal Resource Conservation and Recovery Act and the Clean Water Act have both been violated over the years by dumping that had compromised the health of nearby residents and threatened the drinking water of millions" (Moapa Paiute 2013). The lawsuit alleged that for several years the power plant illegally dumped toxins into the Muddy River, flowing into the Lake Mead reservoir behind Hoover Dam, a source of drinking water for more than 2 million people.

"We are all looking forward to the retirement of the Reid Gardner coal-fired plant that has for decades polluted our reservation," said Vickie Simmons, a leader of the Moapa Band of Paiutes' committee for health and the environment. "And for the sake of our families' health, we must ensure that the toxic waste from the power plant is fully cleaned up. The safety of our community and the future of our children depend on it" (Moapa Paiute 2013). "Now, we have to find out what kind of remediation they're going to do — a complete restoration, a conversion to gas or some other type of project," Tribal President William Anderson told the Associated Press. "To us, the ultimate goal would be to remove everything and put the land back the way it was. We'll be able to come to closure after almost 50 years" (Moapa Paiute 2013). On Earth Day, 2012, the Moapa Paiute had protested the Reid Gardner Power Plant in Las Vegas with members of the Sierra Club on a 50-mile "Cultural Healing Walk" for three days in 100-degree temperatures.

Reservation residents said that the plant has been a threat to their health for several years. William Anderson, Moapa Band Chairman, said: "The high percentage of thyroid and respiratory problems is a big concern for the tribal members on the reservation.…We need a proper study from air monitoring equipment installed on the reservation to study the emissions we're breathing in. That would help determine what needs to be done for our people's health. We also need more stringent storage conditions for coal ash and a study to be conducted to show the health risks associated with breathing in coal ash" (Moapa Paiute March 2012).

"We asked the E.P.A. to at least study the plant or give us a grant to do the study to see exactly what poisons were coming from there, but we were told we didn't qualify for the study or the grant," said Vernon Lee, a Moapa Paiute. "In order to get a federal study grant, there have to be 500 people affected. There are only about 300 members of the Moapa Band and not all of us live on the reservation."

The Indian Country Today Media Network reported that,

> Air pollution takes the form of toxic coal dust, which tribal members say arrives in giant clouds that send people scurrying indoors, but that's not the only problem they have observed. "There are also several settling ponds for coal ash residue, there are enormous piles of coal that are uncovered, and a huge coal ash landfill that is also uncovered," said Barb Boyle of the Sierra Club, in an article published on the Huffington Post's website, written by Mary Ann Hitt, director of the Sierra Club's Beyond Coal campaign. Boyle said that the Tribe "has borne this burden for decades. It's time to stop." The plant was built in 1965, and the Moapa Paiute say they have witnessed their own standard of living plummet over time. "In my era, we were all healthy people," Aletha Tom, who runs the Moapa school house, [said]: "We didn't have the asthma, thyroid problems, cancer, diabetes, but we have that on our reservation. It's so major now."

Soon, due to the tribe's persistence, local television stations and newspapers brought the plague of ash-borne illness to a broader audience. A video "An Ill Wind: The Secret Threat of Coal Ash," reached people from the tribe's website, reporting that the power station "dumps toxin-laden coal ash, a byproduct of combustion, into landfills that lie just a few hundred yards from the reservation. On windy days, coal-ash dust from the plant billows over the reservation, with clouds so thick that you can see and taste them, tribal members say. At such times, residents don't dare let their children play outside. That apparently offers limited protection, though, as the dust seeps into homes, schools and cars". The Sierra Club, Earthjustice, and Greenpeace assisted.

In their long campaign, the Moapa Paiutes won allies. Nevada Senator Harry Reid called the Reid Gardner Power Station a "dirty relic" and supported its closure. Tribal Chairman William Anderson told Indian Country Today Media Network that "The Interior Department gave the Moapa Paiutes fast-tracked approval to build the first-ever utility-scale solar-energy project on tribal lands—which seems especially suitable in this region, with its many months of scorching sun".

The 350-megawatt plant, built on Moapa Paiute trust land, was built to generate enough power for 100,000 homes, according to the agency, which says the project would provide lease income for the tribe, as well as new jobs. "We should have about 400 jobs at peak construction and 15 to 20 permanent jobs—real career jobs tribal members can look forward to," said Anderson. "Our energy customers will likely be in California, where people have an interest in renewable-energy sources.". Construction was planned over several years, along with creation of a preserve for endangered species of desert tortoise. "We're right by an energy corridor, with many above- and below-ground lines for electricity, natural gas, fiber optics, and more, Anderson said. "Connecting is much less expensive than if we were farther away".

As bright as the future may have been, the Moapa Paiute in 2014 were still suffering the coal plant's effects. "Every home has someone—or even everyone—using a breathing apparatus or inhaler," Anderson said. "We see frequent deaths, the most recent being someone in the home closest to the plant". All along, NV Energy said it was continuing its commitment "to operate the Reid Gardner Generating Station in an environmentally responsible manner, in compliance with all federal and state laws, and in the best interests of its customers".

During May 2014, the Moapa were awarded a second solar-energy array by the U.S. Department of the Interior, just as constriction began on the first. The second array, on 850 acres of Moapa land, was sited 20 miles northeast of Las Vegas; the first project is about 30 miles due north of the city. The second array, which will provide power for about 60,000 homes, was constructed to minimize water usage because of persistent drought. Secretary of the Interior Sally Jewell said that the array reflected the Obama Administration's steadfast commitment to work with Indian Country leaders to promote strong, prosperous and resilient tribal economies and communities....This solar project and other grants also delivered on the President's Climate Action Plan goals "to spur important investments and jobs in tribal communities that can be leveraged to address some of the impacts from climate change that threaten tribal lands, waters and ways of life" (Moapa Paiute 2014). By

2020, both plants were finished and operating, with homes and businesses being hooked up. The old coal-fired plant was underway for decommissioning, having belched its last puff of coal dust.

8.12 Montana's Gros Ventre and Assiniboine: Gold Mining and Cyanide Poisoning

The Little Rocky Mountains of Montana, which long have been regarded as sacred by the Assiniboine and Gros Ventre, are now laced by the effluent of open-pit gold mines which have produced toxic acid-mine drainage. Andrew Schneider of the Seattle Post-Intelligencer described Gus Helgeson, the president of Island Mountain Protectors, a Native American environmental and cultural organization, standing atop Spirit Mountain as he scanned "the gashes, pits and piles of rock that once was his tribe's most sacred land" (Schneider 2001). "The strong man weeps," Schneider wrote.

During 1855, the Assiniboine and Gros Ventre were moved to the Fort Belknap Reservation, which was named by the immigrants for a U.S. Secretary of War. The Assiniboine and Gros Ventre gave up 40,000 acres of land in exchange for a government promise to feed, clothe, and care for them. At the time, federal Indian agents said nothing about the gold that was buried in Spirit Mountain, but they made it clear the tribes could either agree to their terms or starve. Spirit Mountain is part of the Little Rockies, an island of mountains in the nearly flat prairie. To Native Americans, the mountains were valued for their deer, bighorn sheep, herbs, natural medicines, and pure water. Gold mining has destroyed all of that.

In 1884, Pike Landusky and Pete Zortman discovered gold on the reservation. Facing starvation, in 1895, the Assiniboine and Gros Ventre signed an agreement negotiated by Gen. George Bird Grinnell, selling portions of their gold-laced land to the federal government for $9 an acre worth of livestock and other goods, as miners besieged towns that had been named after the gold's discoverers. Under the terms of the General Mining Law of 1872, the government sold the land to individuals and private companies for $10 an acre.

A century and a half later, the land was exhausted and broken. Sammy Fretwell wrote in The State (Fretwell 2014): "Sheared off mountaintops, towering piles of rubble and deep pits make it hard to ignore Montana's recent history of gold mining. Dominant on the landscape, industrial-scale gold mines provided jobs and tax revenues for parts of three decades in small communities that came to depend on the economic support. But big open-pit gold mines had such an impact on the environment that Montana effectively banned new ones 16 years ago."

Some of the gold mines exposed sulfide-rich rocks. Once unearthed, whereas sulfide interacted with air and water, sulfuric acid was created that, left by itself, could have leached into the water table for several centuries. A 2006 consulting report found that "60 percent of the hard rock mines researchers examined in the

West had degraded the quality of groundwater and surface water nearby" (Fretwell 2014). Leaching began during the 1980s, and continued about a decade and a half before environmental problems compelled the state to close it in 1998.

The Zortman-Landusky mine was only one of several heap leach mines near the Fort Belknap Indian Reservation. According to the Environmental Justice Atlas (2014):

"This specific mining operation had over a dozen cyanide spills including one incident that resulted in over 50,000 gallons of cyanide being spilled. The mine was also found to be leaking acids, arsenic and lead. This large and frequent contamination led to extensive surface and groundwater contamination" (Zortman-Landusky 2014).

8.13 "Like Watching Our Ancestors Die"

"The first time the mining company let me up here, let me see what they had done to our land, the pain surrounded me. It was like watching our ancestors die, raped of their honor," said Helgeson. "They destroyed this place, took their gold off in armored trucks and left us a wounded mountain spewing poison on the people the mountain was stolen from," he said (Schneider 2001).

Over several decades, according to Schneider's account, "scores of shafts were driven into the Little Rockies, and an estimated $1 billion in gold and silver was taken out of the ground—more than $300 million by the last owner of the mine, Pegasus Gold Corp. of Canada" (Schneider 2001). Underground mining of gold and silver continued until into the 1950s; after that open-pit strip mining began. As gold prices increased rapidly in the late 1970s, the Pegasus Gold Corp. and a subsidiary, Zortman Mining Inc., built mines that extracted gold from heaps of low-grade ore with cyanide solutions. (For one day, the price of gold rose to more than $800 per troy ounce, exceeding the Dow Jones Industrial Average for one day.) The Pegasus Gold Company, owner of several mines in the Little Rocky Mountains, went bankrupt as gold fell sharply after 1980, "leaving the state of Montana with a $100 million cleanup liability and the tribes with the prospect of perpetually polluted water" (Huff 2000). Cyanide-assisted gold mining was a boon to gold-mining companies, even as it portended ecological disaster. Zortman continued mining until 1990, during which time this mine was expanded nine times without any substantial environmental oversight, despite cyanide spills into the water table used by the Indians (Abel 1997).

By 1990, the Assiniboine and Gros Ventre had begun a substantial challenge to the environmental damage wrought by cyanide-heap gold mining, as they formed a Native American environmental-advocacy group, Red Thunder, which joined with other non-Indian environmental groups that resisted federal permits for the Zortman-Landusky's mine's next request for expansion. The group's appeal was denied, however.

8.14 The Mine Leaks, and Expands

During July 1993, a severe thunderstorm sent a large flood of acidified mine wastewater into the town of Zortman. After this flood, the Bureau of Land Management (BLM) required that the mine's owners develop a new reclamation plan. At about the same time, an Environmental Protection Agency study found that the mine had been "leaking acids, cyanide, arsenic and lead from each of its seven drainages" (Abel 1997).

The State of Montana soon joined the E.P.A. in a suit based on the Clean Water Act, which was settled out of court in July 1996, with Pegasus and Zortman Mining pledging to pay $4.7 million in fines to the tribes, the federal government, and the state. The mine's owners also pledged to follow a detailed pollution-control plan in the future. Trusting mine regulators shortly thereafter, approved a request to triple the mine's size (from 400 to 1192 acres), which was approved by the Montana Department of Environmental Quality and the B.L.M. (Abel 1997). The Fort Belknap Community Council and the National Wildlife Information Center in January 1997, sued the Montana Department of Environmental Quality, alleging that the agency's decision to allow an expanded mine violated state law.

During September 1997, Federal and state environmental agencies fined Pegasus and Zortman Mining $25,300 for violating the clean-water settlement by polluting a stream in the Little Rockies the previous summer. John Pearson, director of investor relations for Pegasus, asserted that discharges were the result of "acts of God" during "extraordinarily heavy rains" (Abel 1997). By late 1997, with gold prices (and its share price) declining rapidly, Pegasus warned that its mine would close by January 1, 1998 if the expansion plan was not accepted. The State of Montana and Native American environmental activists wondered whether Pegasus would survive long enough as a corporate entity to complete promised reclamation of existing mines. "Acts of God" aside, In January 1998, Pegasus filed for Chap. 11 bankruptcy protection.

In the meantime, Pegasus left behind open pits that were described by Schneider:

> Pegasus dug pits the size of football fields and lined them with plastic or clay. Crushed ore was dumped in mounds as high as 15 feet and soaked with a mist of cyanide. It became the largest cyanide heap–leach operation in the world.... The heavily contaminated water trickled and flowed through fissures in the mountain, into the surface streams and underground aquifers that supply drinking water for 1000 people who live in and around Lodge Pole and Hays, reservation towns north of the mountains (Schneider 2001).

8.15 Streams Smell of Rotten Eggs

Streams that flowed down the mountain smelled of rotten eggs (the chemical signature of sulfide), cloudy and lifeless. "This is death," said John Allen, a tribal spiritual leader, as he filled his hands with putrid muck. "The mines take millions in gold from our land and leave us poisoned water. The miners and the government experts have argued for years about whether the water is bad. All they have to do is look, but they choose not to see" (Schneider 2001). Allen, 46 years of age in 2001, had thyroid problems, as did three other members of his family. His father also had lymphatic cancer. Doctors who specialize in environmental medicine then told the Assiniboine and Gros Ventre that the diseases afflicting them also stemmed from contaminated water. Environmental advocates among the two tribes also reported an unusually high number of stillbirths.

Health problems plagued people on the reservation long after the company had filed for bankruptcy and escaped its responsibilities by moving out of Montana, leaving "without proper reclamation of the land including leaving toxins like arsenic, hills of waste rock and exposed mountain sides" (Zortman-Landusky 2014). In 1993, the EPA filed a Clean Water Act suit that required a $32 million cleanup, after leaching became obvious. According to one report (Fretwell 2014), "By 1993, selenium, a metal associated with mine waste, began to show up in German Gulch, where an almost pure strain of native cutthroat trout lives.

Tests later revealed that the creek's fish and water bugs were contaminated with selenium, which can hurt the ability of cutthroat trout to reproduce. German Gulch Creek remains unsuitable for aquatic life and for drinking water." Acid drainage from the Zortman and Landusky mines turned some streams orange. Members of both tribes have generally refrained from to drinking water from the area, said Ina Nez Perce, who manages environmental issues for both tribes at the Fort Belknap reservation (Fretwell 2014).

The Gros Ventre and Assiniboine Tribes sued the United States Bureau of Land Management, citing lack of respect for their land from the heap leaching in the Zortman-Landusky mine. Pollutants cited in the suit included cyanide, arsenic, cadmium, aluminum, and iron. In 2005, the state of Montana's Gov. Brian Schweitzer signed legislation providing $1.5 million per year to cleanse contaminated water until 2018.

By 2014, state and federal taxpayers had spent at least $40 million to clean up environmental problems caused by four gold strip mines that shut down in the 1990s, according to the Montana Department of Environmental Quality and the U.S. Forest Service. The Zortman and Landusky mines had cost almost $24 million of that, with several million dollars yet to be expended (Fretwell 2014).

8.16 The Huichol (Wixáritari), a Sacred Site, and Silver Mining

The Huichol (Wixáritari) reside near silver deposits in the Cerro Quemado, a sacred site adjacent to their reserve in Mexico, "the mountain called birthplace of the sun [which] rises up out of the land shrouded by green shrubs and cactuses and towers over the natural marvel that is this landscape. The Huicholes have been making an annual pilgrimage to the Cerro Quemado from their native lands some 250 miles to the west for as long as anyone can remember" (Zhorov 2012). Silver also has been mined there since the eighteenth century. The mines closed in 1991 and Real de Catorce, one of the larger towns in the area, began making money as a tourist destination. The Huichol have (not always willingly) become the object of a New Age invasion by peyote-ingesting Norte Americanos. Tourism declined with rising narco-violence in Mexico, however, and Real de Catorce's population declined to less than a thousand people.

By 2012, with silver prices rising, the miners returned. Minera Real Bonanza (MRB), a subsidiary of the Canadian First Majestic Silver Corp., in a project it calls La Luz ("The Light"), an expanded version of the old mining system. Its object was $3 billion worth of silver, to be extracted from ore. Locals, while desiring jobs that mining may bring, also feared that a large-scale mine would drive away tourists. The real object of controversy was the nature of mining, which has changed. Older, labor-intensive subsurface mining may be replaced by an open pit, as labor costs rise and the ratio of waste rock to silver ore rises. This has been true throughout the mining industry worldwide Strip mining kills the ecosystem, but is more "cost-effective" in the world of accounting.

Royalties often are not paid to local people in Mexico, and open-pit mining may ruin the land and water. Once the life of a mine ends (usually about 15 years), local people may be living in a wasteland. The MRB, however, promised 500 direct jobs that would generate 1500 more indirectly after work began. The company also asserted that it would mine discretely, (whatever "discretely," might mean regarding a strip mine), that "mine entrances and tailings will be tucked into the landscape and invisible to tourists. . .The company also promised a water-treatment plant—Real's first—to treat "black water" instead of ejecting it directly into streams, as previously. "The company has prepared a closing plan, which was said to involve total remediation of the site" (Zhorov 2012). The company also proposed an exhibition mine for tourists, and a mining museum.

Some of the company's concessions are a result of activism by the Zapatista uprising that made indigenous rights a political issue in Mexico. In 2008, the Hauxa Manaka Pact for the Preservation and Development of the Huichol Culture was signed to protect Huicholes' sacred places and culture generally. The Defense Front of Wirikuta held educational forums and fund-raising concerts drawing as many as 50,000 people to appreciate the sacred nature of Cerro Quemado and other aspects of their peyote culture. The mining company promised that Cerro Quemado would not be strip-mined.

"In the company's offices hung a piece of Huichol art depicting a slice of Huichol mythology, in psychedelic colors and dreamscapes," reported Irina Zhorov of the Indian Country Today Media Network. However, not everyone trusted the company's promises. María Teresa Sánchez Salazar, who teaches in the geography department of the National Autonomous University of Mexico, said that throughout the country promises and laws are not the problem. "The issue is for them to actually abide by the legislation," she says. "There isn't supervision strict enough to make them honor their obligations" (Zhorov 2012).

References

"360 Scientists Urge Environmental Protection Agency to Quash Bristol Bay Pebble Mine". Indian Country Today Media Network. February14, 2014. http://indiancountrytodaymedianetwork.com/2014/02/14/360-scientists-urge-environmental-protection-agency-quash-bristol-bay-pebble-mine-153559. Last accessed May 24, 2015.

Abel, Heather. "The Rise and Fall of a Gold Mining Company." High Country News 29:24 (December 22, 1997). http://www.hcn.org/servlets/hcn.Article?article_id=3860. Last accessed October 11, 2000.

"Anglo American Withdraws from Pebble Mine". Environment News Service, September 20, 2013. http://ens-newswire.com/2013/09/20/anglo-american-withdraws-from-alaskas-pebble-mine/. Last accessed August 22, 2014. Last accessed October 13, 2013.

"Bristol Bay Tribes' Fight to Fend off Pebble Mine Highlighted in National Geographic". Indian Country Today Media Network, November 19, 2012. http://indiancountrytodaymedianetwork.com/bristol-bay-tribes-fight-to-fend-off-pebble-mine-highlighted-in-national-geographic. Last accessed May 18, 2012.

Chythlook-Sifsof, Callan J. "Native Alaska, Under Threat." New York Times, June 28, 2013. http://www.nytimes.com/2013/06/28/opinion/native-culture-under-threat.html. Last accessed January 30, 2014.

"Company That Designed Burst B.C. Tailings Pond Was Hired by Pebble Mine in Bristol Bay". Indian Country Today Media Network, August 8, 2014. http://indiancountrytodaymedianetwork.com/2014/08/08/company-designed-burst-bc-tailings-pond-was-hired-pebble-mine-bristol-bay-156315. Last accessed April 3, 2017.

Cortes, Amber. "Treaty Rights and Totem Poles: How One Tribe Is Carving Out a Resistance to Coal. http://exp.grist.org/lummi Grist: A Beacon in the Smog. No date. Last accessed August 27, 2014.

"Dam Breach at Mount Polley Mine in British Columbia." NASA Earth Observatory". August 17, 2014. Last accessed September 1, 2015. http://earthobservatory.nasa.gov/IOTD/view.php?id=84202&src=eoa-iotd

Eagland, Nick. "Deep Divide Between Anti- and Pro-pipeline Rallies in Metro Vancouver," Vancouver Sun, March 19, 2018. http://vancouversun.com/news/local-news/live-hundreds-protest-kinder-morgan-pipeline-expansion-in-burnaby. Last accessed May 2, 2022.

Fretwell, Sammy. "Toxic Legacy Haunts Montana; What Can SC [South Carolina] The State (Charleston, S.C.), October 11, 2014. http://www.thestate.com/news/local/article13897079.html. Last accessed January 31, 2015.

Gentry, Don and Emma Marrismarch. "The Next Standing Rock? A Pipeline Battle Looms in Oregon." New York Times, March 8, 2018, no page.

Green, Sara Jean. "Fighting a GIANT." Windspeaker, no date. Accessed October 22, 2005. Last accessed November 30, 2005. http://www.ammsa.com/classroom/CLASS3Lubicon.html

Guerette, Deb. "No Clear-cut Answer: Timber Rights Allocation on Lubicon Land a Worrisome Development." Grande Prairie Daily Herald-Tribune, March 5, 2001. http://www.tao.ca/~fol/Pa/negp/ht010305.htm. Last accessed April 16, 2001.

Hansen, Terri. "Northwest Tribes Fight for Treaty Rights in Face of Coal-Transport Plan." Indian Country Today Media Network, August 15, 2012.

Hansen, Terri. "Major Environmental Disasters in Indian Country." Indian Country Today Media Network, October 8, 2013. http://indiancountrytodaymedianetwork.com/2013/10/08/7-major-industrial-environmental-disasters-indian-country-151661. Last accessed November 10, 2015.

"Horrific Toxic Spill in B.C. Called Another Exxon Valdez". Indian Country Today Media Network. August 7, 2014. http://www.nativenewstoday.com/2014/08/07/horrific-toxic-spill-in-b-c-called-another-exxon-valdez/. Last accessed March 23, 2016.

Huff, Andrew. "Gold Mining Threatens Communities." The Progressive Media Project. July 11, 2000. http://www.progressive.org/mpdvah00.htm. Last accessed October 3, 2000.

Jones, Brendan. "A Gold Rush in Salmon Country." New York Times, November 24, 2017. https://www.nytimes.com/2017/11/24/opinion/sunday/gold-mine-salmon.html. Last accessed January 22, 2018.

LeBlanc, Judith. "Lead with Love." *Together We Rise: The Women's March: Behind the Scenes at the Protest Heard Around the World.* New York: HarperCollins, 2018.

"Lubicon Chief Collected $1.5M, While Community Had No Running Water: Audit". Canadian Broadcasting Corporation (CBC), September 12, 2014. http://www.cbc.ca/news/indigenous/lubicon-chief-collected-1-5m-while-community-had-no-running-water-audit-1.2763972. Last accessed December 3, 2015.

"Lummi Nation Officially Opposes Coal Export Terminal in Letter to Army Corps of Engineers". Indian Country Today Media Network, August 2, 2012. http://indiancountrytodaymedianetwork.com/2013/08/02/lummi-nation-officially-opposes-coal-export-terminal-letter-army-corps-engineers-150718. Last accessed October 3, 2012.

Mapes, Lynda V. "'Like Standing Rock': Trans Mountain Pipeline-Expansion Opponents Plan B.C. Protest." Seattle Times, March 8, 2018. https://www.seattletimes.com/seattle-news/environment/like-standing-rock-trans-mountain-pipeline-expansion-opponents-plan-b-c-protest/?utm. Last accessed March 22, 2018.

McNeel, Jack. "Railroad to Disaster? Inland Tribes Fight to Avert Coal-Train Destruction." Indian Country Today Media Network, December 17, 2012.

"Moapa Paiute to Host Second Solar Project on Its Lands". Indian Country Today Media Network, May 29, 2014. http://indiancountrytodaymedianetwork.com/2014/05/29/moapa-paiute-host-2nd-solar-project-its-lands-155071. Last accessed July 11, 2014.

"Moapa Paiute March 50 Miles in Anti-Coal Protest". Indian Country Today Media Network, April 27, 2012. Last accessed July 15, 2012. http://indiancountrytodaymedianetwork.com/article/moapa-paiute-march-50-miles-in-anti-coal-protest-110450

"Moapa Paiute Sue Over Coal Plant Contaminants". Indian Country Today Media Network, August 9, 2013. Last accessed November 14, 2013. http://indiancountrytodaymedianetwork.com/2013/08/09/moapa-paiute-sue-over-coal-plant-contaminants-150806. Last accessed October 22, 2013.

"Native Alaskans Laud Environmental Protection Agency's Nixing of Pebble Mine in Bristol Bay". Indian Country Today Media Network, March 3, 2014. Last accessed September 8. 2015. http://indiancountrytodaymedianetwork.com/2014/03/03/native-alaskans-laud-environmental-protection-agencys-nixing-pebble-mine-bristol-bay. Last accessed September 5, 2001.

"Northwest Tribes Step Up Opposition to Proposed Coal Terminals". Indian Country Today Media Network, October 19, 2012. Last accessed October 30, 2012. http://indiancountrytodaymedianetwork.com/article/northwest-tribes-step-up-opposition-to-proposed-coal-terminals-140853

Oil Spill in Lubicon Lake Cree Nation, Alberta: KAIROS urgent action May 6, 2011. KAIROS Canadian Ecumenical Justice Initiatives. The Forum on Religion and Ecology at Yale. http://

fore.yale.edu/news/item/oil-spill-in-lubicon-lake-cree-nation-alberta-kairos-urgent-action/. Last accessed June 18, 2011.

"PBS News Hour: Lummi Fighting Cherry Point Coal Export Terminal". Indian Country Today Media Network, August 4, 2012. Last accessed October 22, 2012. http://indiancountrytodaymedianetwork.com/2013/08/04/pbs-newshour-highlights-lummi-battle-over-cherry-point-coal-export-terminal-150726. Last accessed November 13, 2012.

"Pour Him a Tall One! Mining Exec Insists He'd Drink Water from Tailings Pond". Indian Country Today Media Network, August 7, 2014. Last accessed August 23, 2014. http://indiancountrytodaymedianetwork.com/2014/08/07/pour-him-tall-one-mining-exec-insists-hed-drink-water-tailings-pond-156294

"Quapaw Take Lead in Tar Creek Superfund Site Cleanup in Oklahoma, Kansas". Indian Country Today Media Network, June 4, 2014. Last accessed October 5, 2014. http://indiancountrytodaymedianetwork.com/2014/06/04/quapaw-take-lead-tar-creek-superfund-site-cleanup-oklahoma-kansas-155155

"Resisting Destruction: Chronology of the Lubicon Crees' Struggle to Survive". No date. Last accessed October 23, 2017. http://www.lubiconsolidarity.ca/resisting.html

Riordan, Michael. "Don't Sell Cheap U.S. Coal to Asia." New York Times, February 13, 2013, A-23. Last accessed April 3, 2013. http://www.nytimes.com/2014/02/13/opinion/dont-sell-cheap-us-coal-to-asia.html

Schneider, Andrew. 'A Wounded Mountain Spewing Poison:' The Mining of the West: Profit and Pollution on Public Lands." *Seattle* Post-Intelligencer, June 12, 2001. http://seattlepi.nwsource.com/specials/mining/27076_lodgepole12.shtml. Last accessed July 30, 2001.

Shephard, Dan. "Last Residents of Picher, Oklahoma Won't Give Up the Ghost (Town)." NBC News, April 28, 2014. http://www.nbcnews.com/news/investigations/last-residents-picher-oklahoma-won-t-give-ghost-town-n89611. Last accessed April 28, 2014.

Trahant, Mark. "Elections 2012: Vote for Coal? Northwest Tribes Raise New Concerns." Indian Country Today Media Network, October 1, 2012. Last accessed December 4, 2012. http://indiancountrytodaymedianetwork.com/article/elections-2012%3A-vote-for-coal%3F-northwest-tribes-raise-new-concerns-137150

"Treaty Victory as Northwest Tribes Celebrate Oregon Coal Train Rejection". Indian Country Today Media Network, August 21, 2014. Last accessed September 2, 2014. http://indiancountrytodaymedianetwork.com/2014/08/21/treaty-victory-northwest-tribes-celebrate-oregon-coal-train-rejection-156520

"Unions Back Lubicons". Indigenous Environmental Network. August 16, 2000. http://www.ienearth.org/lubicon.html#canada. Last accessed November 8, 2000.

Virji, Anar. "Shadows of a Town Destroyed by Toxic Waste." Aljazeera. May 23, 2014. http://www.aljazeera.com/blogs/americas/2014/04/98746.html. Last accessed July 5, 2014.

Walker, Richard. "Lummi Totem Pole Journey Rallies Voices Against Environmental Destruction." Indian Country Today Media Network, September 2, 2014. Last accessed November 16, 2014. http://indiancountrytodaymedianetwork.com/2014/09/02/lummi-totem-pole-journey-rallies-voices-against-environmental-destruction-156696

Walker, Richard. "Lummi Call Coal Terminals an Absolute No-Go, Invoking Treaty Rights." Indian Country Today Media Network, February 26, 2015. Last accessed March 30, 2015. http://indiancountrytodaymedianetwork.com/2015/02/26/lummi-call-coal-terminals-absolute-no-go-invoking-treaty-rights-159381

Warrick, Joby. "'Gold vs. Salmon: EPA Becomes Target by Employing Rare Preemptive 'Veto.'" Washington Post, February 15, 2015. Last accessed May 2, 2015. http://www.washingtonpost.com/national/health-science/internal-memos-spur-accusations-of-bias-as-epa-moves-to-block-gold-mine/2015/02/15/3ff101c0-b2ba-11e4-854b-a38d13486ba1_story.html

Zhorov, Irina. "Reopening Silver Mine Could Save Town, but Destroy the Sacred Site of the Huicholes." Indian Country Today Media Network, August 7, 2012. Last accessed October 2, 2012. http://indiancountrytodaymedianetwork.com/article/reopening-silver-mine-could-save-town,-but-destroy-the-sacred-site-of-the-huicholes-127,755.

Zortman-Landusky. Gold Mine, Montana, USA. Environmental Justice Atlas. 2014. https://ejatlas.org/conflict/gold-mining-in-montana.Last accessed December 4, 2014.